Inequality and Climate Change

Inégalité et changement climatique

This book is a product of the South-South Tricontinental Collaborative Programme of CLASCO, CODESRIA and IDEAs.

Ce livre est une compilation d'articles issus du programme tricontinental Sud-Sud entre CLASCO, CODESRIA et IDEAs.

Inequality and Climate Change
Perspectives from the South

Inégalité et changement climatique
Perspectives du Sud

Editd by / Sous la direction de

Gian Carlo Delgado-Ramos

Consejo Latinoamericano
de Ciencias Sociales
Conselho Latino-americano
de Ciências Sociais
CLACSO

CODESRIA

© CODESRIA 2015
Council for the Development of Social Science Research in Africa
Avenue Cheikh Anta Diop, Angle Canal IV
BP 3304 Dakar, 18524, Senegal
Website : www.codesria.org

ISBN : 978-2-86978-645-5

Typesetting: Alpha Ousmane Dia
Cover Design: Ibrahima Fofana

Distributed in Africa by CODESRIA
Distributed elsewhere by African Books Collective, Oxford, UK
Website: www.africanbookscollective.com

The Council for the Development of Social Science Research in Africa (CODESRIA) is
an independent organisation whose principal objectives are to facilitate research, promote
research-based publishing and create multiple forums geared towards the exchange of views
and information among African researchers. All these are aimed at reducing the fragmentation
of research in the continent through the creation of thematic research networks that cut across
linguistic and regional boundaries.

CODESRIA publishes *Africa Development*, the longest standing Africa based social science
journal; *Afrika Zamani*, a journal of history; the *African Sociological Review*; the *African Journal
of International Affairs*; *Africa Review of Books* and the *Journal of Higher Education in Africa*.
The Council also co-publishes the *Africa Media Review*; *Identity, Culture and Politics: An
Afro-Asian Dialogue*; *The African Anthropologist, Journal of African Tranformation, Method(e)
s: African Review of Social Sciences Methodology*, and the *Afro-Arab Selections for Social Sciences*.
The results of its research and other activities are also disseminated through its Working Paper
Series, Green Book Series, Monograph Series, Book Series, Policy Briefs and the CODESRIA
Bulletin. Select CODESRIA publications are also accessible online at www.codesria.org.

CODESRIA would like to express its gratitude to the Swedish International Development
Cooperation Agency (SIDA), the International Development Research Centre (IDRC), the
Ford Foundation, the Carnegie Corporation of New York (CCNY), the Norwegian Agency
for Development Cooperation (NORAD), the Danish Agency for International Development
(DANIDA), the Netherlands Ministry of Foreign Affairs, the Rockefeller Foundation,
the Open Society Foundations (OSFs), TrustAfrica, UNESCO, UN Women, the African
Capacity Building Foundation (ACBF) and the Government of Senegal for supporting its
research, training and publication programmes.

Contents /Sommaire

List of Figures, Tables and Map

Figures

Tables

Map

Contributors / Les auteurs

Tara Caetano completed her master's degree in energy studies in 2011, with a thesis on the socioeconomic impacts of South Africa's electricity development plan (Integrated Resource Plan 2010). At the time of writing, she was based at University of Cape Town's Energy Research Centre. Prior to this, she was at the German Institute of Global and Area Studies in Hamburg, where her work focused on researching win-win situations for emissions reduction and development in the South African context. She also spent a couple of months at United Nations University World Institute for Development Economics Research (UNU-WIDER) in Helsinki on a research stay working with James Thurlow, developing a Computable General Equilibrium model for South Africa. Tara's research interests are primarily on pursuing a development-first approach to climate change. She is particularly interested in finding synergies between mitigation, employment creation and poverty alleviation.

Adrien Coly est enseignant chercheur à l'Université Gaston Berger de Saint Louis. Docteur en géographie de l'Université Cheikh Anta Diop de Dakar, il est certifié en gestion intégrée des ressources hydriques pour les pays en voie de développement de la Fondation Universitaire Luxembourgeoise en Belgique. Il est membre du conseil de Laboratoire « *Leidi* » – *dynamique des milieux et développement* – et dirige le « pole eau » dénommé « Gouvernance des territoires de l'eau » dont les travaux de recherche s'intéressent au rapport de l'eau au territoire suivant une approche « syndrome ». Dans le domaine des risques naturels et vulnérabilités (risques urbains, sécurité en eau, pollution des milieux), Dr Coly coordonne différents programmes de recherche à l'Université Gaston Berger et il est l'auteur de plusieurs articles et publications scientifiques.

Saudamini Das is an associate professor in Environment and Development issues at the Institute of Economic Growth, Delhi (on lien from Swami Shradhanand College of University of Delhi). Saudamini is a Fellow of South Asian Network for Development and Environmental Economics (SANDEE) and has also worked as Mälar scholar at the Beijer Institute of Ecological Economics,

Royal Swedish Academy of Sciences, Stockholm. Her research areas are natural disaster analysis, valuation of ecosystem services, coastal vulnerability, adaptation to climatic extreme events (storms and heat waves), and evaluation of public policy among others. Saudamini has publications in reputed journals such as *Proceedings of National Academy of Sciences, USA; Climate Change Economics; and Natural Hazard; Estuarine, Coastal and Shelf Sciences; Economic and Political Weekly.* Her significant research on storm protection services of mangroves during the October 1999 super cyclone in India has been turned into as a short movie by the American Museum of Natural History and has also been published as a special article; "The Mystery of Mangroves" by *Nature Conservancy* magazine in its 'Summer 2010' issue.

Gaston Fulquet is a PhD candidate in the Social Sciences Doctoral Programme of *Facultad Latinoamericana de Ciencias Sociales* (FLACSO Argentina), assistant lecturer and coordinator for the Global Studies Programme at the same institution. In addition he works as consultant with the National Environmental Secretariat of Argentina. His areas of interest include global governance, regional and interregional cooperation, international political economy, emerging powers and environmental politics.

Félicien Kabamba Mbambu, est de la République démocratique du Congo, il est titulaire d'un doctorat en Sciences Politiques et Administratives de l'Université de Kinshasa où il est professeur de Sciences Politiques au Département des Sciences Politiques et chercheur au Centre d'Etudes Politiques. Auteur de nombreuses publications, ses recherches sont orientées vers les questions des ressources naturelles (Forêts, mines, agriculture) en Afrique centrale avec un accent sur les thématiques de REDD+ et de climat.

Ranjan Kumar Mohanty is a PhD Candidate, Centre for Economic Studies & Planning, School of Social Sciences, Jawaharlal Nehru University, New Delhi, India.

Claudine Valérie Ouédraogo épouse Rouamba est Maître assistante au département de Sociologie à l'Université de Ouagadougou. Elle possède un doctorat en Sciences de l'éducation obtenu à l'Université Paris 8 Vincennes-St Denis et un DESS en Sociologie du développement de l'Institut d'Etudes du Développement Economique et Social IEDES-Paris 1 Panthéon Sorbonne. Actuellement chargée de cours à l'Université de Ouagadougou, de Koudougou et catholique St Thomas D'Aquin, elle est membre du laboratoire Genre et

Développement et du Groupe de recherche sur les initiatives locales (GRIL). Dr Ouédraogo est également membre de la Chaire « Femmes, Sciences, Sociétés et Développement Durable » de l'Université de Ouagadougou et Vice présidente de la Commission Nationale des Droits Humains du Burkina Faso. Elle effectue des recherches sur les politiques éducatives (innovations pédagogiques et reformes) et sur la problématique de l'inégalité de genre.

Gian Carlo Delgado-Ramos graduated with a degree in economics from the National Autonomous University of Mexico (UNAM), before proceeding to the Autonomous University of Barcelona, Spain, for his master's and doctoral studies in Ecological Economics, Environmental Management, and Environmental Sciences, . He is a full time researcher on "City, Management, Land and Environment" program of the Interdisciplinary Research Centre on Sciences and Humanities at UNAM. He is also member of the National Research System of the National Science and Technology Council (CONACYT), Mexico. He is the recipient of the Mexican Academy of Sciences Research Prize on Social Sciences – 2014 and the National University Prize for Young Researchers – 2011. He is the lead-author of Chapter 12, Group III, of the 5th Assessment Report of the Intergovernmental Panel on Climate Change (IPCC).

Fatimatou Sall est assistante scientifique et pédagogique à l'Institut Africaine de Gestion Urbaine dans le cadre d'un master en « aménagement urbain et environnement » et doctorante au Laboratoire LEIDI/Pôle Gouvernance des Territoires de l'Eau de l'université Gaston Berger de Saint-Louis (Sénégal). Elle fait une thèse unique sur « La résilience au prisme de la ville et des écosystèmes : analyse de la dynamique des zones humides et ses impacts sur les services écosystémiques dans la ville de Saint-Louis. Géographe environnementaliste, Fatimatou est également titulaire d'un master professionnel en *Développement spécialité Gestion des Aires Protégées* de l'université Senghor d'Alexandrie (Egypte). Elle a participé au programme de recherche dénommé CLUVA (Climate change and Urban Vulnerability in Africa) en tant que chercheur junior de Mars 2011 à Décembre 2013.

Natéwindé Sawadogo is a sociologist with a PhD on Science, Technology and Society, University of Nottingham. He is member of Laboratoire de Recherche Interdisciplinaire en Sciences Sociales et Santé, Department of Sociology, University of Ouagadougou, Burkina Faso. His area of interest include Medical sociology, Science and Technology Studies and Social theory.

James Thurlow is an applied economics scholar, whose research focuses on the interactions between policies, economic growth and poverty in low-income countries. Before joining the United Nations University World Institute for Development Economics Research (UNU-WIDER), he was a Research Fellow at the International Food Policy Research Institute, where he researched into rural and regional development strategies, agricultural investments, and weather risk and climate change. He holds a PhD in economics from the University of Natal (South Africa), and has worked throughout eastern and southern Africa, and in Bangladesh, Ghana, Peru and Vietnam. His research interests include agriculture and rural development, climate change adaptation and mitigation, economic growth and poverty, and economic modeling techniques.

Gabriela Canedo Vásquez is a sociologist. She graduated from Universidad Mayor de San Simon (UMSS), Bolivia and later obtained her master's and PhD degrees on Social Anthropology from the Center for Research and Advanced Studies in Social Anthropology, Mexico. . She currently teaches at the UMSS. From the year 2011 to 2015, she was a UNDP consultant, and served as vice-president of the Bolivian Studies Association from 2013 to 2015. Her research work and publications focus on territorial disputes, indigenous people, gender, democracy and multiculturalism. Some recent publications are: "La Loma Santa una utopia cercada. Estado, territorio y cultura en la Amazonia boliviana" (2011); and "Diez años SAN-TCO. La lucha por los derechos territoriales indígenas de Tierras Bajas de Bolivia" (2008).

Nirmala Velan is faculty member and Chair of the Deptment of Economics, Pondicherry University, India. She specializes in gender studies, labor markets, and development economics. She has published papers in refereed national/ international journals and she is author of several books. She is a member of various academic committees and examination boards.

Introduction

Gian Carlo Delgado-Ramos

Climate change, an anthropogenic phenomenon caused by the emission of cumulative greenhouse gases (GHG), mostly from fossil fuel combustion (see Figure 1)[1] has become a ponderous issue, as attested by many unprecedented changes over time (IPCC 2014a). As the Intergovernmental Panel on Climate Change (IPCC) said, the last three decades has been successively warmer on the Earth's surface than any preceding decade since 1850, while the period from 1983 to 2012 was *likely* the warmest 30-year interval of the last 1,400 years in the Northern Hemisphere (ibid). As a result, the global average temperature has increased by 0.85oC since 1880 and it is expected to further rise by 1.5oC by the end of the century in relation to the average temperature of the 1850–1900 period (ibid).

Figure 0.1: GHG emissions by Economic Sectors

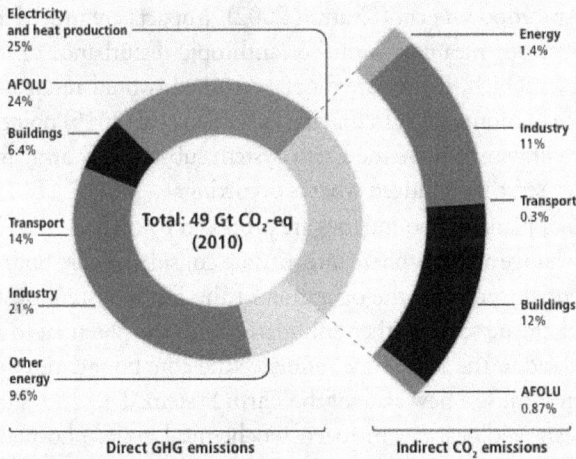

Besides the increase in oceanic temperature (measured to a depth of 75m) by about 0.11oC per decade since 1971, the absorption of increasing loads of CO2 by oceans has resulted in their surface acidification: pH has decreased by 0.1, corresponding to a 26 per cent increase in acidity, a change that has negative implications for marine biodiversity, mainly on coral reefs (ibid). Additionally, global sea level has risen since

1901 by 0.19m, while the Greenland and Antarctic ice sheets, as well as glaciers, have shrunk (ibid). Cumulative emissions between 1750 and 2011 are estimated at 2,040 ±310 GtCO2eq, of which 40 per cent still remains in the atmosphere. Despite a growing number of climate change mitigation policies, states the 5th Assessment Report of the IPCC, annual GHG emissions grew on average by 1.0 GtCO2-eq (2.2 per cent) per year, from 2000 to 2010, compared to 0.4 GtCO2eq (1.3 per cent) per year, from 1970 to 2000 (ibid). And, since half of total anthropogenic emissions occurred in the last 40 years, economic and population growth are now acknowledged, without doubt, as the main drivers of climate change (ibid).

The correlation of economic growth and the use of energy and materials by human societies during the last century is clear cut: while human population increased fourfold and the global economy about 14 times, material and energy use increased tenfold on average. Biomass use increased 3.5 times, energy use 12 times, metal ores 19 times, and construction minerals, mainly cement, about 34 times (Krausmann et al 2009). By the end of the XX century, humanity used about 500 thousand petajoules of primary energy and about 60 billion tons of raw materials yearly (Weisz and Steinberger 2010). Unequal use is however significant, as the highest consuming 10 per cent of the world population uses 40 per cent and 27 per cent of the world's energy and materials respectively (ibid).

Accordingly, human transformation of nature is of such magnitude that biochemical cycles of the planet are being eroded, a context in which extreme climate events are only a partial manifestation.

Observed human impacts on nature have been documented and categorized by some as the Anthropocene era (Crutzen 2002). Impacts are indeed a transgression of *planetary boundaries*, meaning 'limits' of anthropic disturbance of the planet Earth's critical processes which, if they were not perturbed, would result in a relatively safe operating space for human life. In that sense, Steffen et al (2015) point out sensibly that '…it would be unwise to drive the Earth System substantially away from a Holocene-like condition'. Yet, that is indeed what is occurring.

Two levels of planetary boundaries are proposed by Steffen et al (2015). On a first level, climate change and biosphere integrity are considered core boundaries since they have the potential to change the operation of the Earth System on their own. On a second level, there are several other boundaries with the potential to affect the quality of human life and at the same time influence the core boundaries; however, on their own they cannot cause a new state of the Earth System. These are stratospheric ozone depletion, ocean acidification, nitrogen biochemical cycle, phosphorus biochemical cycle, land-system change, human use of fresh water, atmospheric aerosol loading and the introduction of novel entities (such as chemical pollution). Table 1 summarises the main features of such planetary boundaries as Rockström *et al* 2009 and Steffen *et al* 2015 as described them) and their current state.

Table 0.1: Planetary Boundaries

Planetary boundary	State before 1850	Proposed boundary		Current state
		Rockström et al. 2009	Steffen et al. 2015	
Climate change	280 parts per million	<350 parts per million	<350-540 parts per million	396.5 parts per million
			Energy imbalance +1.0Wm-2	2.3 Wm-2
Change in biosphere integrity		Loss of biodiversity (10 species per million)	Genetic diversity (10 species per million, with a proposed goal of 1 per million)	100 species per million
			Functionality of diversity (90% intact biodiversity index)	84% (based on Southern Africa only)
Stratospheric ozone depletion	290 DUs***	276 DUs	<5% reduction from preindustrial level of 290 DUs (5% - 10%) assessed by latitude.	283 DU (Rockström et al 2009); only transgressed over Antarctica during southern hemisphere springtime (~200 DUs; Steffen et al 2015)
Ocean acidification**	3.44 arag**	2.75 arag**	≥80%-≥70% of preindustrial aragonite saturation state of ocean surface average.	290 arag (Rockström et al 2009); about 84% of the preindustrial aragonite saturation state (Steffen et al 2009)
Nitrogen biochemical cycle	0 tons per year	35 million tons per year	62 Tg N per year-1	121 million tons/year (Rockström et al. 2009); about 150 Tg N/year-1 (Steffen et al 2015)
Phosphorus biochemical cycle	1 million tons per year	11 million tons per year	Global cycle not greater than 11 Tg P per year-1	8.5 - 9.5 million tons per year (Rockström et al. 2009); about 22 Tg P per year-1 for the global cycle and about 14 Tg P per year-1 for the regional cycle (Steffen et al 2015)
			Regional cycle not greater than 6.2 Tg P per year-1	

Land-system change	Low	15%	Area of forested land as % of original forest cover on a global scale (75-54%); and area of forested land as % of potential forest as part of a biome (tropical: 85-60%; temperate: 50-30%; boreal 85-60%)	11.7% (Rockström et al 2009); 62% (Steffen et al 2015)
Human use of freshwater (alteration of water cycle)	415 cubic kilometres	4,000 cubic kilometres yearly1	Global use of 4,000 cubic kilometres yearly1 and monthly withdrawal no greater than 25-55% at basin level in low-flow months; 30-60% in intermediate flow-months, and 55-85% in high-flow months.	2,600 cubic kilometres yearly1
Atmospheric aerosol loading	----	----	Global Aerosol Optic Depth (AOD). AOD as seasonal average for a given region (Study case, monsoons in South Asia)	0.30 AOD in the Southern Asian region.
Introduction of novel entities	Non-existent	Unknown****		Unknown****

Source: Compiled by the author, based on Rockström et al, 2009, 'Planetary Boundaries: Exploring the Safe Operating Space For Humanity', *Ecology and Society*, Vol. 14. No. 2. Article 32; Steffen et al, 2015, 'Planetary Boundaries: Guiding Human Development on a Changing Planet', Scienceexpress. DOI: 10.1126/science.1259855.

Notes: *It is estimated that, as from 1751, 337 billion tons of carbon have been emitted, exclusively by burning fossil fuels.

**A reduction in the value means an increase in acidification. Figures represent the state of aragonite saturation.

***A Dobson Unit, or DU, is the equivalent of 0,01 mm. depth of the ozone layer under normal pressure and temperature conditions.

****There are no indicators that might enable us to measure this type of pollution in a standardized way, although there are some methodological proposals for specific toxic substances. Some of the substances singled out are persistent organic polluting substances, plastics, endocrine disruptors, heavy metals, radioactive waste and nanomaterials.

The transgression of planetary boundaries, starting with climate change, has profound implications for practically all biophysical and human systems. Food production, water resources, land and costal ecosystems, as well as human health are causes for special concern due to their particular sensitivity and implications for the resilience of nature and life quality of future generations. Other implications derived from the above could be related to the exacerbation of existing challenges such as land tenure insecurity, poverty and inequality (including gender issues), marginalization of poorer (principally rural) populations, climate induced migration, and resource wars or conflicts.

As suitably expressed by Klein, '…the thing about a crisis this big is that it changes everything. It changes what we can do, what we can hope for, what we can demand from ourselves and our leaders. It means there is a whole lot of stuff that we have been told is inevitable that simply cannot stand. And it means that a whole lot of stuff we have been told is impossible has to start happening right away' (Klein 2014:28). This is doubly true in the case of developing countries where climate change implications will be felt more keenly because of biophysical reasons, but also because inequality and lack of governance are particularly challenging issues, remarkably in low-income countries.

Considering that both, biophysical and socioeconomic issues determine human vulnerability, it can then be said that the poorest population already suffers and will certainly experience most of climate change impacts. In words of the IPCC, '…climate change will amplify existing risks and create new risks for natural and human systems. Risks are unevenly distributed and are generally greater for disadvantaged people and communities in countries at all levels of development' (IPCC 2014a:13). Yet, climate change implications may be phased down through climate policy and corresponding commitments and actions.

Thus, an approach from the global South for adaptation and mitigation of climate change, as the one offered by this book, is certainly necessary and appropriate for enhanced future options, preparedness, integrated responses and cooperation, chiefly when GHG contributions are asymmetrical. For example, Africa contributes less than 4 per cent of global GHG emissions while Europe, with a smaller population, is responsible for more than three times that amount (about 13 per cent). Furthermore, a differentiated historical responsibility for climate change has been acknowledged:[2] while Annex I Parties[3] have positive historical differentiated responsibilities, non-Annex I Parties have negative historical differentiated responsibilities in relation to their contribution to climate change, and this doesn't account for indirect emissions related to imports which are relevant as high-income countries are importing large embodied emissions from the rest of the world, mainly the upper middle-income countries whose GHG emissions have risen steadily over the last decade with substantial differences between mean and median per capita emissions (in this case, China, now the greatest global GHG emitter, serves as an example) (IPCC 2014b).

Since GHG emissions keep rising in a context of only half-successful mitigation, the risk of abrupt or irreversible changes increases, as it is correlated to the magnitude of global warming. Bearing in mind that many aspects of climate change and its associated impacts will continue for centuries, even if anthropogenic GHG emissions are checked, international climate negotiations have set a goal to ensure that the increase in temperature is not greater than $2°$ C (relative to the 1861-1880 period). This requires, however, that accumulated CO2eq emissions since 1870 don't exceed 2,900 Gt of CO2eq, in a context in which it is to be noted that at the end of 2011 about 1,900 Gt of CO2eq had already been released (IPCC 2014a).

The future of climate depends then on both historical and future emissions and the absorption capacity of carbon sinks (plants, soils and oceans), apart from natural climate variability. Therefore, in order to tackle climate change, more efforts are needed at multi-scale levels (international, regional, national and subnational), including a more comprehensive long-term and integrated vision, political will and suitable and democratic governance structures, funding, effective decision-making and capacity to respond, as well as further articulated actions that will differ across sectors and regions. As the IPPC puts it: '...*adaptation and mitigation responses are underpinned by common enabling factors. These include effective institutions and governance, innovation and investments in environmentally sound technologies and infrastructure, sustainable livelihoods and behavioural and lifestyle choices.*' (ibid: 26).

While it is true that mitigation actions are needed from all parties, it is without doubt a specific requirement to major emitters, or all Annex-I countries and the more economically dynamic upper-middle income countries. Adaptation actions, which should better be implemented from a comprehensive and integrated climate change (adaptation-mitigation) agenda, are of particular relevance to low-middle income countries since socioeconomically induced climate vulnerability is higher among the poorer population, as has already been explained.

In addition to the above, one may plead for challenging current power relationships and inequalities while upholding human rights as a starting point for the construction of genuine alternative development pathway(s) capable of transcending socioeconomic inequalities and embracing cultural diversity, while keeping the planet under a Holocene-like condition.

Taking the above into account, and from a global South point of view, academic (and non-academic) inquiries on the multifaceted nature of climate change become necessary and appropriate, including those analysing socioeconomic, political and cultural aspects. This was one of the main goals of the Comparative Research Workshop on 'Inequality and Climate Change: Perspectives from the South' of the South-South Collaborative Programme of CLACSO-CODESRIA-IDEAS, celebrated on 24 and 25 July 2014 in Dakar, Senegal. Additionally, it aimed to identify common spheres for cooperation amongst the countries of the South, as well as to address any shared perspective to enhance informed participation in global debates on climate change.

This book is an outcome of this workshop, and certainly a fresh contribution in the context of preparations for the 21st Climate Change Conference to be held in Paris by the end of 2015.

The papers included have been organised geographically, starting with those dealing with African case studies, followed by those from Latin America and Asia. Authors' contributions are widely multi- and interdisciplinary, and embrace a plurality in their appraisals on a set of issues, from renewable energy, climate induced migration, gender and poverty, to the use of traditional pharmacopeia in a climate change context, international cooperation for biofuel production, water-energy nexus and water disputes in urban settlements, flooding disasters and their social implications, and the impact of heat waves on workers.

The ample biophysical differences from one case study to the other, as well as their particular social and cultural realities, make this book invaluable and unique, as it offers a vantage point from which to examine some of the current perspectives on inequality and climate change from the global South.

August 2015

Notes

1. It is estimated that 78 per cent of anthropogenic GHG emissions since 1970 are due to fossil fuels combustion and industrial processes (IPCC 2014a). Of the total source of greenhouse gases at present, CO_2 contributes 76 per cent; CH_4 about 16 per cent, N_2O about 6 per cent and the combined fluorinated gases (F-gases) about 2 per cent (IPCC 2014b).

2. From 1850 to 1960, GHG emissions increased due to a large industrialisation process, particularly in the US but also in Germany and UK; however, by the 1950s, China and Russia started seeing their emissions climb as well (still behind the top emitters mentioned before, particularly the US). From 1960, the US kept its place as the top GHG emitter until 2005 when China emerged as the top world GHG emitter (other Asian emergent economies saw their emissions climb as well in comparison to their own historical trajectory). Therefore, one may conclude that there is a historically differentiated responsibility from those countries that have experienced a major and sustained industrialization process since 1850, such as the US, Germany and UK. In this context, reticence from China to commit to a major GHG reduction effort is associated to such historical responsibility, as it has been argued that such climate commitments may impose limits to its recent economic development. The truth is that developed countries have already spent most of their carbon budget.

3. According to the Framework Convention on Climate Change, Annex I Parties include the industrialized countries that were members of the OECD (Organisation for Economic Co-operation and Development) in 1992, plus countries with economies

in transition (the EIT Parties), including the Russian Federation, the Baltic States, and several Central and Eastern European States. Non-Annex I Parties are mostly developing countries (http://unfccc.int/parties_and_observers/items/2704.php).

References

Crutzen, Paul, 2002, 'Geology of Mankind'. *Nature*. No. 415: 23.

IPCC, 2014a, Climate Change 2014. Synthesis Report. *Contribution of Working Groups I, II and III to the Fifth Assessment Report of the Intergovernmental Panel on Climate Change,* Geneva :IPCC.

IPCC, 2014b, Introductory Chapter. In: *Climate Change 2014: Mitigation of Climate Change. Contribution of Working Group III to the Fifth Assessment Report of the Intergovernmental Panel on Climate Change,* Cambridge: Cambridge University Press.

Klein, Naomi, 2014, *This Changes Everything. Capitalism vs. The Climate.* New York: Simon & Schuster.

Krausmann, *et al.*, 2009, 'Growth in Global Material Use, GDP and Population during the 20th Century', *Ecological Economics,* No. 68, pp. 2696 – 2705.

Rockström, J., *et al.*, 2009, 'Planetary Boundaries: Exploring the Safe Operating Space for Humanity', *Ecology and Society,* Vol. 14, No. 2. Available at: www. ecologyandsociety.org/vol14/iss2/art32/

Steffen, *et al.*, 2015, 'Planetary Boundaries: Guiding Human Development on a Changing Planet', *Sciencexpress,* DOI: 10.1126/science.1259855.

Weisz, Helga, and Steinberg, Julia, 2010, 'Reducing Energy and Material Flows in Cities,' *Environmental Sustainability,* Vol. 2, pp. 185.

1

The Socioeconomic Implications of Renewable Energy and Low Carbon Trajectories in South Africa[1]

Tara Caetano and James Thurlow

Introduction

South Africa is highly coal dependent with a large variance between emissions per capita and levels of development. The current structure of the South African economy has resulted in sub-optimal outcomes: environmentally with high carbon intensity and socially with a Gini-coefficient of 0,63[2] 29.2 per cent of the population living on US$2.5 a day[3] and an official unemployment rate of 24.3 per cent.[4] High levels of poverty and inequality are likely to be exacerbated substantially by climate change impacts in the future.[5]

South Africa has committed to emissions reduction of 34 per cent by 2020 and 42 per cent by 2025 relative to a 'business-as-usual' baseline (RSA 2010). In order to reach these targets alternative energy options need to be explored. The country's Integrated Resource Plan (IRP) shows a move in the right direction with a decrease in the reliance on coal-fired plants and an increase in renewable energy generation capacity.

The current process of the IRP is influenced by a number of policy goals, including emissions reductions. These policy goals act as 'inputs' into the operational process. The intention of the IRP is to address these and propose an electricity supply plan that is aligned with these policy goals and ensures the supply of affordable and reliable electricity to the region. Three easily quantifiable indicators form the basis of decision making in the IRP; namely investment cost, emissions reduction and water usage. There are, however, a number of important economic and social policy goals that should also form an integral part of the decision making process, namely: (1) economic growth or GDP growth; (2) employment; (3) regional development; (4) localisation; (5) good terms of trade; and (6) low electricity price. The modelling approach used in the IRP is limiting in terms of analysing the plan's ability to address some of these policy goals. This is a major gap in the planning process, as these policy goals are important considerations for economic growth and development nationally as well as regionally. An interim attempt was made during the IRP process to quantify the possible effects of scenarios on these policy goals. The process followed a Multi-Criteria Decision Making[6] methodology informed by various stakeholder meetings. An important drawback of this method is that it is difficult to prove that there is solid theoretical backing for the results and that these results are not influenced by subjectivity. However, under time and budget constraints it was difficult to include a thorough economic analysis in the IRP process, and the need for this type of analysis was mentioned in the draft report for the IRP (DoE 2010).

This paper aims to fill this gap in the literature by using a highly disaggregated economy-wide model to analyse the potential socioeconomic implications of introducing renewable energy and implementing a carbon tax in South Africa. Furthermore, it seeks to use the model to address the impacts on two of the policy goals in the IRP, namely, economic growth and employment. The chosen methodology is appropriate for the analysis as it is theory-based and consistent with the current structure of the South African economy.[7]

There are a few existing studies that use similar methodologies to simulate mitigation actions in South Africa. Pauw (2007), Devarajan et al. (2011) and Alton et al. (2012) explore issues surrounding a carbon tax in South Africa. Devarajan et al. (2011) find that the implementation of a carbon tax in South Africa is likely to lead to a decrease in welfare but is, however, more efficient than other tax instruments in curbing energy use and emissions. An important limitation of this study, highlighted in Alton et al. (2012), is that there is no differentiation between energy technologies or inclusion of the country's long-term electricity investment plan. Pauw (2007), on the other-hand, distinguishes between different types of energy technologies and uses a partial-equilibrium energy model[8] to derive an optimal electricity investment schedule. This study finds smaller welfare reductions from the introduction of a carbon tax in comparison to Devarajan et al. (2011). Alton et al. (2012) follow Pauw (2007) by including detailed energy technologies and deriving electricity investment paths from an energy sector model. Secondly, they address a number of limitations of the aforementioned studies: the use of a dynamic CGE to overcome the lack of time dimension; industries are allowed to invest in less energy-intensive activities in response to higher energy prices; labour and capital market rigidities are captured; a number of tax recycling options are simulated. A carbon tax of R12 per ton of CO_2 is introduced in 2012 and projected to rise linearly to a value of

R210 per ton in 2022; sufficient to meet the national emissions reduction target. This study highlights the importance of both the design of the carbon tax as well as the method of revenue recycling. In comparison, the use of tax revenues to fund corporate tax reductions is favourable for economic growth and high-income households but results in decreased welfare for the majority of the population. An alternative option of expanding social transfers, intuitively, improves welfare for low-income households but results in less economic growth.

The methodology used in this paper follows on from that used in Alton *et al.* (2012). The model design is extended to include a highly disaggregated renewable energy sector. Three scenarios, based on scenarios derived from a partial equilibrium energy sector model[9] used in the IRP process are simulated in this paper. The scenarios depict different levels of renewable energy investment and, since they are derived from an energy model, are consistent with South Africa's electricity system requirements. The results will include a comparison between potential impacts of these scenarios on economic growth, inequality, employment and emissions reduction.

Electricity Generation Options in South Africa

Description of the Model Scenarios

The Integrated Resource Plan (IRP) broadly describes the process of modelling and decision making for the future of South Africa's electricity generation. The main objectives are to, first, estimate the long-term future demand for electricity and, second, to identify possible scenarios of generation capacity that are able to meet this demand (DoE 2011). The long lead times and high investment costs associated with electricity generation capacity provide obvious motivation for the importance of integrated resource planning. A number of other concerns accompany these in the case of South Africa; economic uncertainty due to the long time horizon, pending emissions reduction targets, and security of supply concerns due to the country's dominant reliance on coal, to name but a few (DoE 2011).

The scope of the IRP spans over the total demand and supply for electricity in South Africa, and includes Eskom as well as non-Eskom sources of generation capacity. The foundation of the plan is built on a number of policy recommendations, such as cost-minimisation, emissions constraints, regional development and localisation potential.

The initial stage of the IRP requires the generation of a base case, or reference scenario. This base case represents the least cost option and is considered the optimal option in terms of meeting capacity needs when the only limitation is the cost factor (DoE 2011). There are a number of other scenarios that are then compiled in the light of explicit policy and the consideration of risk adjustments that eventually lead to the determination of a proposed electricity build plan for South Africa.

A number of policy requirements govern the IRP. These form the foundation on which the IRP is built. Three particular elements of policy are crucial to the determination of the plan. Firstly, the Energy White Paper (DME 1998) specified a preference for the movement away from reliance on coal and towards a more diverse electricity generation mix with the inclusion of nuclear, natural gas and renewable options. Secondly, in light of potential future international climate change obligations the IRP is considerate of South Africa's climate change policy. With regard to this, the importance of accounting for the environmental impacts of electricity generation technologies is noted and should be accounted for in the IRP. Thirdly, much political pressure is applied to ensure that electricity provision remains at the least possible cost to the consumer. In light of this, the purpose of the IRP is to provide a capacity build plan to meet the expected demand growth at the minimum social cost. The cost should include the costs associated with the impact of externalities.

The ultimate goal of the IRP process is to present a build plan that is acceptable to the Ministry as the most optimal scenario depicting a number of constraints and policy interests. The plan is not 'set in stone' and should be revised every two years in an attempt to mitigate the effects of uncertainty and allow the plan to evolve to meet revised demand growth and include any technological developments that may occur over the period. The current scenario is the policy-adjusted plan; considered to be a compromise between the least cost scenario (base case) and the scenario with the strictest emissions target. But it is also the most costly, being the emissions 3 scenario. The use of these three scenarios in this paper allows an appropriate contrast between employment projections under a low carbon trajectory and under a 'business-as-usual trajectory', where there is no need to reduce emissions. Figure 1 provides a graphical representation of the total new capacity builds under these scenarios over the period of analysis, 2010 to 2030.

Figure 1.1: The Planned Capacity Builds for all Three Scenarios (GW)

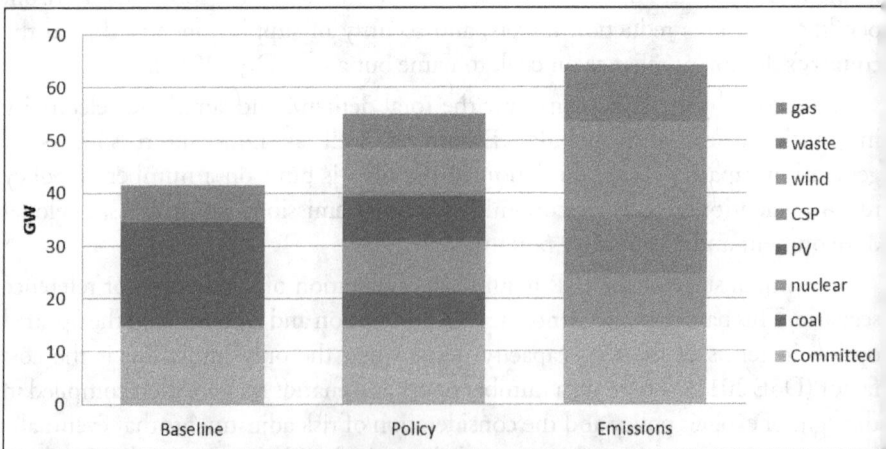

Source: Own calculations based on the IRP (2011)

The least-cost technology option in South Africa is coal, with coal-fired plants supplying over 90 per cent of South Africa's electricity. This is apparent in the baseline scenario where capacity for coal-fired electricity generation almost doubles over the period to 2030. There are a number of capacity build plans that are considered 'firm commitments' and are either in the process of being built or in the final stages of planning. Two large coal-fired plants, Medupi and Kusile, make up the bulk of the committed builds and are planned to add 8,760 MW of capacity by 2020 (dependent on delays). A number of small renewable electricity generation plants are also considered 'committed', however their contribution is minor in comparison, with an estimated 2,400 MW by 2030.

The policy-adjusted scenario displays a more diversified electricity build plan, with the inclusion of 9600MW of nuclear power, and 8400 MW each of wind and solar photovoltaic (PV). There is also an increase in peaking capacity, open-cycle gas turbines (OCGT) and closed-cycle gas turbines (CCGT), with 6280 MW of capacity in total. The emissions 3 scenario relies heavily on the use of renewable energy, contributing to approximately 60 per cent of total electricity capacity by 2030. As in the policy-adjusted scenario, 9600MW of nuclear power is planned to come online during the period with no additional base-load capacity from coal-fired plants. The emissions reductions in this scenario, although substantial with an annual emissions limit of 220MT CO2-eq, still won't get South Africa to the targeted emissions reduction of 42 per cent from baseline by 2025. Alton *et al.* (2012) estimate that, given domestic demand forecasts and production quotas, at least an additional R0.46 trillion of investment would be needed for South Africa to reach its emissions reduction target. The emissions pathways for the three scenarios are given below:

Figure 1.2: Emissions Pathways for the Base Case, Policy-adjusted and Emissions 3 Scenarios

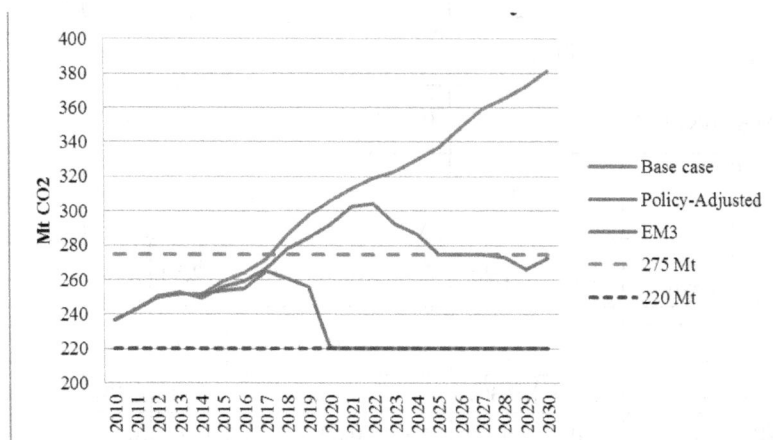

Source: Based on IRP calculations

In order to ensure that the scenarios are comparable, we simulate the same total electricity supply in GWh for all scenarios. Renewable options for electricity generation currently have low capacity factors, in comparison to nuclear power and coal-fired plants. The rest of this section will expand on the technology options available in the IRP.

Technology Options for Electricity Generation

There are some alternative electricity generation options outlined in the IRP. Each option producing the same good, electricity, but with different technology coefficients – i.e., they have different factor and intermediate inputs. Table 1 provides a summary of the technology options in terms of cost, demand for intermediates and factor demand.

Table 1.1: Intermediate and Factor Estimates for Electricity Generation Technologies

	Coal	Nuclear	Hydro	PV	CSP	Wind	Waste	Gas	Diesel
Base Year 2007									
Electricity Supply (GWh)	229 571	11 317	5 845	213	319	32	204	1	86
Gross Operating Surplus10 (R mil)	55 749	2 480	1 369	140	103	8	76	0	16
Total Employment (people)	33 014	2 071	2 063	64	96	7	56	0	12
High Skilled (people)	15 054	795	990	32	48	3	26	0	6
Assumptions[11]									
Build cost (Rmil/GWh)	17 785	26 575	9 464	37 225	37 425	14 445	9 464	4 868	4 868
Levelised Cost12 of Plant (Rmil/GWh)	0,40	0,74	0,13	1,43	1,42	0,70	0,54	0,96	2,25
O&M (jobs/GWh)	0,14	0,18	0,35	0,30	0,30	0,22	0,27	0,14	0,14
Construction/ Installation (Job years/MW)	10,40	10,80	19,40	52,30	10,80	4,50	6,90	6,20	6,20
Manufacturing (Job years/MW)	1,50	1,20	0,90	16,80	7,20	22,50	0,80	0,07	0,07
Imported Content (%)	35	35	35	70	50	70	50	35	35
Value[13] (R/GWh)	6 225	9 301	3 312	26 058	18 713	10 112	4 732	1 704	1 704
Fuel (Rmil/GWh)	0,08	0,07	0,00	0,00	0,00	0,00	0,00	0,60	2,39

Source: Own calculations based on EPRI (2010)[14]

Measuring Economy-wide Impacts

Structure of South African Economy and Labour Markets

Table 2 outlines the structure of the South African economy and labour market in 2007. South Africa has a dominantly services-based economy, with services accounting for over 66 per cent of gross domestic product (GDP) and approximately two-thirds of employment. The electricity sector is a relatively small sector, with a contribution of around 1.8 per cent of GDP and 0.3 per cent of employment. Historically cheap electricity prices coupled with a mineral-rich country has aided the development of energy-intensive sectors in the economy. For this reason, we believe that the importance of the electricity sector is understated when looking at the direct contribution to GDP; the indirect effects of changes in the electricity sector are more pronounced given the forward linkages associated with the sector.

Table 1.2: Structure of South Africa's Economy and Labour Market (in percentage)

| | Share of Total | | | | Exports/ Output | Imports/ Output |
	GDP	Employment	Exports	Imports		
Total GDP	100,00	100,00	100,00	100,00	11,21	15,28
Agriculture	3,11	3,74	2,64	0,95	11,14	5,65
Industry	30,77	29,08	83,73	84,22	21,49	27,53
Mining	8,83	8,79	33,41	10,47	65,07	40,75
Coal Mining	1,59	1,61	4,49	0,21	43,82	4,31
Manufacturing	16,83	15,88	48,75	72,47	16,55	30,04
Petroleum	1,15	0,20	2,17	3,67	8,41	21,84
Electricity	1,81	0,31	1,57	1,29	15,22	14,43
Coal-fired	1,63	0,28	-	-	-	-
Nuclear	0,15	0,02	-	-	-	-
Hydro	0,02	0,01	-	-	-	-
Construction	2,70	3,93	-	-	-	-
Services	66,12	67,18	13,63	14,83	3,11	3,91

Source: South Africa 2007 social accounting matrix (own calculations)

Eskom is the state utility and runs a monopoly in the electricity sector, generating approximately 95 per cent of the electricity used in South Africa and an estimated 45 per cent of the electricity used in Africa (Eskom 2011). Electricity generation is highly reliant on the use of coal, which remains the cheapest generation option given that South Africa is a coal-rich country. There was not much diversity in terms of electricity generation in 2007, with approximately 93 per cent of electricity generated by coal-fired plant; 1.8GW (5 per cent) generated by Koeberg, Africa's first nuclear power station; and the remainder generated mainly from hydropower (Eskom 2011).

Description of the Static E-SAGE Model

A number of CGE models have contributed to the local policy making process in areas including trade strategy, income distribution, and structural change in low-income countries. There are several features of this class of models that make them suitable for this type of analysis (Arndt, Davies, & Thurlow 2011). Firstly, the structure of CGE models ensures that all economy-wide constraints are respected and provide a theoretically consistent framework for welfare and distributional analysis (Arndt, Davies, & Thurlow 2011). Secondly, CGE models simulate the functioning of a market economy, and provide a platform for analysis on how different economic conditions affect markets and prices (Arndt, Davies, & Thurlow 2011). One of the drawbacks of this type of modelling, however, is that the credibility of the results is highly dependent on the accuracy of the data and assumptions made when calibrating the model. It is possible to mitigate this limitation through transparency and disclosing the assumptions made and data used in building the economy-wide model.

The South African General Equilibrium (SAGE) model used in this analysis is derived from neoclassical tradition originally presented in the seminal work by Dervis, de Melo, & Robinson (1982). A number of extensions and adaptations have been made to this framework including the ability for producers to produce more than one commodity and the explicit treatment of transaction costs (Lofgren, Harris, & Robinson 2001). The dynamic-recursive energy extension to the SAGE model, developed by Channing Arndt, Rob Davies and James Thurlow (2011) is used in this paper. The SAGE model was extended to reflect the detailed structure and workings of South Africa's energy sector. In addition, the model was developed further to capture a detailed factor demand for the electricity sector. The SAGE model is a dynamic recursive model; in simple terms a sequence of static model runs that are solved to simulate the passing of time. The static model is solved 'within-the-period' with the use of non-linear equations that are solved simultaneously to capture linkages that exist in the real economy. This is followed by a 'between-period' run where a number of parameters are updated according to exogenous behavioural changes over time as well as the results from the previous static run. The E-SAGE model simulates the period between 2010 and 2030 and each static run represents one year.

There are 46 productive sectors, or *activities*, identified within the model; as well as six factors of production including, capital, crop land and labour. Labour is disaggregated further into four factors by level of education – primary, middle, secondary, tertiary.

The production schedule for a sole producer is provided for simplicity, although, in reality, the SAGE model contains 46 sectors, each of which are assigned a representative producer. The behaviour of the representative producer is such that they will maximise profits subject to a given set of input and output prices (Thurlow 2004). The model follows neoclassical theory, and assumes constant returns to scale and, hence, a constant elasticity of substitution (CES) function is used to determine production (Arndt, Davies, & Thurlow 2011):

$$QA_i = \alpha_i^p \left(\Sigma_f \delta_{if}^p \cdot QF_{if}^{-p_i^p} \right)^{-1/p_i^p}$$

(1)

where QA is the output quantity of sector i, α^p is the shift parameter reflecting total factor productivity (TFP), QF is the quantity demanded of each factor f (i.e., labour and capital) and δ^p is a share parameter of factor f employed in the production of good i. The elasticity of substitution between factors σ is a transformation of ρ^p.

The use of a CES function allows producers to respond to changes in relative factor returns by smoothly substituting between available factors to derive a final value-added composite (Thurlow 2004).

Profits π in each sector i are defined as the difference between revenues and total factor payments (Arndt, Davies, & Thurlow 2011):

$$\pi_i = PV_i \cdot QA_i - \Sigma_f(WF_f \cdot QF_{if})$$

(2)

where PV is the value-added component of the producer price, and WF is factor prices (e.g., labour wages and returns on capital). Profit maximisation implies that factors will receive an income where marginal revenue is equal to marginal cost, based on endogenous relative prices (Thurlow 2004). Maximising sectoral profits subject to Equation 6, and rearranging the resulting first order condition provides the system of factor demand equations used in the model (Arndt, Davies, & Thurlow 2011):

$$QF_{if} = \alpha_i^{\frac{p_i^p}{1+p_i^p}} \cdot QA_i \left(\delta_{if}^p \cdot \frac{PV_i}{WF_f} \right)^{1/(1+p_i^p)}$$

(3)

According to Arndt *et al.* (2011), the SAGE model assumes a Leontief specification for technology when calculating the intermediate demands of individual goods as well as when merging aggregate factor and intermediate inputs. This use of fixed shares is due to the belief that technology, and not the decision making of producers, determines the mixture of intermediates per unit of output, and the ratio of intermediates to value-added (Thurlow 2004). In light of this the complete producer price PA is (Arndt, Davies, & Thurlow 2011):

$$PA_i = PV_i + \Sigma_j(PQ_j \cdot io_{ij})$$

(4)

Where io_{ij} represents the fixed input-output coefficient used in the demand for intermediates, which defines the quantity of good j used in the production of one unit of good i (Arndt, Davies, & Thurlow 2011).

The SAGE model represents an open-economy and hence the model recognise the two-way trade that exists between countries for similar goods (Arndt, Davies & Thurlow 2011). Substitution possibilities, governed by a CET function, exist between the production for domestic and for foreign markets, (Thurlow 2004). A CET function is used to allow the distinction between domestic and imported goods in terms of differences in time and/or quality that may exist between them (Thurlow 2004).

Producers are driven by profit maximisation and therefore choose to sell in the market that offers the highest returns (Thurlow 2004). Exported commodities are disaggregated further using a CES according to the specific region under a CES specification (Thurlow 2004). The assumption that the substitution between regions is governed by a CES specification is fair as one would expect that producers would react to changes in relative prices across regions. This would therefore change the geographical composition of their exports accordingly (Thurlow 2004).

The import market is treated in the same regard. Substitution possibilities exist between imported and domestic goods under a CES Armington specification (Armington 1969). This is true in the use of both final and intermediate goods (Arndt, Davies, & Thurlow 2011).

The SAGE model distinguishes between different institutions that exist in the South African economy; namely, households, government and enterprises. Households are disaggregated according to income deciles, except for the top decile, which is divided into five income categories (Thurlow 2004).

The factor income generated from production forms the primary source of income for households and enterprises (Thurlow 2004). In addition, due to the model representing an open economy, household incomes consist of transfers from the government, other domestic institutions as well as from the rest of the world. Factor returns in South Africa have been found to differ across both occupations and sectors. In this light, the SAGE model uses a fixed activity-specific wage-distortion term combined with the economy-wide wage to generate activity-specific wages that are paid by each activity (Thurlow 2004). There are a number of assumptions governing the factor market. Firstly, the supply of capital is fixed over a specific time-period, i.e. fully employed, but is considered immobile across sectors (Thurlow 2004). Energy capital, however, is treated as fully employed and activity-specific. There is assumed to be unemployment for the unskilled workers, however, the other three labour categories are assumed to be fully employed and mobile. Remittances are also received by factors from the rest of the world and therefore also contribute to factor incomes (Thurlow 2004).

The SAGE model follows general equilibrium theory in that households within a certain income category are assumed to share identical preferences, and are therefore modelled as 'representative consumers' (Thurlow 2004). According to this theory, equilibrium is reached when the representative household maximises their utility subject to a budget constraint. In the model, each representative household has its own utility function, in which QH is the level of consumption is income-independent and constrained by the households' marginal budget share (Arndt, Davies, & Thurlow 2011). Utility is maximised for the consumer subject to a budget constraint, in which PQ is the market price of each good, YH is total household income, and sh and th are marginal savings and direct income tax rates, respectively (Arndt, Davies, & Thurlow 2011). By maximising the above utility function subject to a household budget constraint, a linear expenditure system (LES) of demand is derived (Arndt, Davies, & Thurlow 2011).

The LES of demand represents the consumer preferences captured in the model, given prices and incomes. These demand functions define households' real consumption of each commodity (Thurlow 2004). The LES specification is used in the model as it allows the identification of excess household income and therefore ensures a minimum level of consumption (Thurlow 2004).

The government is considered to be a separate agent with income and expenditure, although it isn't considered to have any behavioural functions (Arndt, Davies, & Thurlow 2011). Most of the income earned by the government is from direct and indirect taxes and its expenditure is assumed to be on consumption and household transfers (i.e., grants) (Thurlow 2004).

Household and enterprise savings are collected into a 'savings pool' from which investment in the economy is financed (Thurlow 2004). It is assumed in the model that government borrowing can diminish this supply of loanable funds and that capital inflows from the rest of the world are able to increase it (Thurlow 2004). There is no specified behavioural function governing the level of investment demand in the model, although the model assumes that the total value of investment spending must equate the total amount of investible funds TI in the economy (Arndt, Davies, & Thurlow 2011).

The SAGE model assumes full employment and factor mobility across sectors. Thus the following factor market equilibrium holds (Arndt, Davies, & Thurlow 2011):

$$\sum_i QF_{if} = QFS_f \tag{5}$$

where QFS is fixed total factor supply. Assuming all factors are owned by households, household income YH is determined by (Arndt, Davies, & Thurlow 2011):

$$YH_h = \sum_{if} \omega_{hf}(1 - tf_f) \cdot WF_f \cdot QF_{if} \tag{6}$$

where $\omega\omega$ is a coefficient matrix determining the distribution of factor earnings to individual households, and tf is the direct tax on factor earnings (e.g., corporate taxes imposed on capital profits).

The model is set up with a number of closures that govern macro adjustments. The selection of appropriate closures should ensure that the model reacts to shocks in a way that is representative of the real economy under investigation. There are considered to be three broad macroeconomic accounts in the SAGE model: the current account, the government balance and the savings and investment account (Thurlow 2004). The macroeconomic balance in the SAGE model is governed by a number of closure rules, which provide a mechanism through which adjustments are made to maintain this balance, or equilibrium (Arndt, Davies, & Thurlow 2011).

According to Arndt, *et al.* (2011), the current account is considered to be the most important of these macro accounts. A substantial amount of research pours into this topic, although in this case due to the single-country open economy CGE model it is considered an exogenous variable (Arndt, Davies, & Thurlow 2011). It is assumed that a flexible exchange rate adjusts in order to maintain a fixed level of foreign borrowing for the current account macro closure rule (Thurlow 2004). South Africa's firm commitment to a flexible exchange rate system and idea that foreign borrowing is unlimited ensure that the chosen closure rule is realistic (Thurlow 2004).

The second closure rule concerns the government balance. The government consumption spending in the SAGE model is considered to be exogenous. In response to this the fiscal balance, or government savings are flexible and adjust accordingly (Arndt, Davies, & Thurlow 2011).

The third closure rule, perhaps the least obvious, involves the choice of a savings-investment closure (Thurlow 2004). The relationship between savings and investment continues to be a highly debated and controversial topic in macroeconomics (Nell 2003). Neo-classical along with new endogenous growth theory maintains the view that it is former savings that decide an economy's investment and output (Thurlow 2004). Conversely, from a Keynesian perspective, it is investment that is exogenous and savings that adjust accordingly (Thurlow 2004). Although, according to Nell (2003), recent works have established that in the case of South Africa, the long-run savings and investment relationship is associated with exogenous savings and no feedback from investment. The SAGE model assumes a balanced savings-driven closure where government and investment expenditure are fixed shares of absorption, determined by a scaled marginal propensity to save (mps).

Along with these three macroeconomic accounts, a factor market closure exists in the model. The various factors in the economy require specification in terms of how they are to be treated in the model. The SAGE model assumes full employment for high-skilled labour and unemployment amongst low-skilled labour with labour being mobile across sectors – a suitable closure for the South African context (Pauw 2007). Capital stock is assumed to be fully employed and activity-specific for the electricity sector, as the simulations impose a structural shift on production capacity. Land is assumed to be fixed and immobile as it is generally treated.

The consumer price index is assumed to be the numeraire in the SAGE model (Arndt, Davies, & Thurlow 2011). In other words, all prices are considered relative to the weighted unit price of household's initial consumption bundle (Arndt, Davies, & Thurlow 2011).

The Energy Sector and Carbon Tax Simulations

Electricity is defined as a single commodity in the SAGE model comprised of each electricity subsector's (i.e. nuclear, hydropower, etc.) separate supply onto the national grid. The model assumes that each of these subsectors has its own distinctive production technology, based on estimates from an earlier study by Pauw (2007). It is also assumed that each subsector requires a different mix of factor inputs (Arndt, Davies, & Thurlow 2011). Hence, there are different electricity 'activities' but a sole electricity commodity. This is a realistic assumption as consumers in South Africa are not able to demand certain 'types' of electricity as it all comes from the national grid; electricity subsectors have very different supply processes and costs.

A number of adjustments had to be made to allow multiple energy subsectors produce the same commodity. The updated production functions are adapted to:

$$QAS_{is} = \alpha_{is}^p \left(\sum_f \delta_{isf}^p \cdot QF_{isf}^{-p_{is}^v} \right)^{-1/p_{is}^v} \tag{7}$$

$$QF_{isf} = \alpha_{is}^{p^{\frac{p_{is}^v}{1+p_{is}^v}}} \cdot QAS_{is} \left(\delta_{isf}^p \cdot \frac{PV_{is}}{WD_{isf} \cdot WF_f} \right)^{1/(1+p_{is}^v)} \tag{8}$$

$$PAS_{is} = PV_{is} + \sum_j PQ_j io_{ijs} \tag{9}$$

where QAS is the output of subsector s within aggregate sector i, PAS is the subsector producer price, and io reflects each subsector's unique production technology. Factor demands QF are also defined at sector level.

A high elasticity of substitution is assumed to exist between energy subsectors in order to replicate their product homogeneity (Arndt, Davies, & Thurlow 2011). However, switching between different energy subsectors is constrained by the fixed installed capital in each subsector, due to the immobility of this capital (Arndt, Davies, & Thurlow 2011). The speed at which South Africa can exchange between energy sources is determined by new capital investment as installed capital is assumed to depreciate at a fixed rate (Arndt, Davies, & Thurlow 2011). In the current extension to the SAGE model, new investment in each subsector is determined exogenously and follows the Integrated Resource Plan (IRP) (Arndt, Davies, & Thurlow 2011).

Energy is treated as an intermediate input in the E-SAGE model; aggregated with other intermediates using a Leontief production function. Producers are, however, able to respond to energy price changes by the use of a 'response' elasticity (ρ). The energy product input coefficient (ioij) falls either when energy prices rise (provided there is some new investment) or when the new investment share (sj) is positive (provided the price rises). This relationship is illustrated below:

$$io \downarrow ij,t + 1/io \downarrow ijt = 1 - (1 - P \downarrow jt/P \downarrow j,t - 1 \uparrow -\rho) \cdot s \downarrow i$$

The carbon tax simulations were applied domestically; similar to an ad valorem tax placed only on fossil fuels burned within the South African borders. We assumed that there was a uniform reduction in indirect sales tax rates to have a less severe, distribution neutral simulation. An important next step would be to model tax recycling options, especially in light of the findings from Alton et al. (2012) that show that the choice of revenue recycling is a main driver of the economic impact of a carbon tax in South Africa. The modelling of alternative recycling options was not conducted in this paper due to time constraints; however, based on the results from Alton et al. (2012) mention will be made of the potential impacts of these alternative options on our results.

The carbon tax design proposed by the National Treasury for South Africa is highly complex (RSA 2013). At first glance, the proposed R120 per ton of CO2 seems to be a significant tax allocation, although it is only half of the carbon tax value estimated by Alton et al. (2012), if South Africa is to reach emissions reduction targets. The Treasury proposed an initial phasing-in period from 2015 to 2019 with the rate increasing at 10 per cent annually until the end of 2019. The rate of increase for the second period, 2020 to 2025, will be announced in February 2019. All sectors will benefit from a 'basic tax-free threshold' of 60 per cent of emissions as well as a number of complex exemptions for energy-intensive users. The electricity sector will benefit from an additional 5 per cent to 10 per cent exemption whilst the petroleum sector will be exempt from an additional 15 per cent to 20 per cent for being a trade-exposed sector. Energy intensive sectors such as chemicals, glass, cement, iron and steel, ceramics and fugitive emissions from coal mining will benefit from exemptions of up to 85 per cent (RSA 2013). The effective tax rate is therefore much lower at between R12 and R48 per ton of CO2, likely to be too little to transform South Africa's emissions pathway.

The carbon tax simulated in this analysis is designed in a more simplistic manner. The carbon tax is also assumed to phase in between 2015 and 2019, increasing linearly over the period until a total of R120 per ton of CO2 is levied on all sectoral emissions. Given that the effective tax rate is significantly lower than this, the scenarios will overestimate the proposed carbon tax. The decision not to include the exemptions is, firstly, to simplify this initial analysis and, secondly, because existing literature suggests that an effective tax rate of between R12 to R48 per ton is not enough to have a significant impact on South Africa's emissions trajectory.

Results and Discussion

The simulations were run under two conditions: one without a carbon tax and a second with a simplified carbon tax. The next step would be to model the exact tax design proposed by the Treasury and compare the socioeconomic implication with this simplified version of the tax; an interesting modelling exercise for the future. As previously noted, alternative revenue recycling options not modelled in this paper is on the agenda for future work.

Table 3 presents the results for the simulations run without a carbon tax. All three scenarios fair quite favourably in terms of growth in South Africa, with a slightly lower average growth rate for the policy-adjusted scenario and more so for the emissions 3 scenario. It should be noted that the assumptions governing the financing of the electricity build plan might be resulting in an overly optimistic economic growth projection. It is assumed that the build plan is financed by a foreign loan, of which an annual interest payment of 5 per cent is made annually; none of the principal payment is made over the modelling period to 2030. This may be a contentious assumption; however, given that economy-wide models are not predictive but rather are a valuable tool for comparing possible futures, the relative burden on the economy should be sufficient for our analysis. It would be interesting to explore different financing options and analyse the potential impacts of these on the economy – a topic that should be noted for future work.

Table 1.3: Simulation Results without a Carbon Tax (in percentage)

	GDP Growth	Inequality	Emissions Reduction	Employment
Base	3,90	1,10	0	1,32
Policy	3,82	1,01	-11	1,31
Emissions 3	3,67	0,85	-18	1,29

Source: Author's calculations

The emissions 3 scenario requires significantly more investment in comparison to the base case and to a lesser extent the policy-adjusted scenario. This is shown in the slight contraction of the economy relative to the base case. Economic growth is still positive, but the higher investment cost results in a decrease in the investment funds available for other, more profitable sectors in the economy.

The second indicator is titled '*inequality*'; in this instance the values refer to the relative increase in income growth for poorest decile in comparison to the riches decile.[15] In the base case, the income of the poorest decile increases by 1,1 per cent over the simulation period, in relation to the richest decile; the income gap is narrowing slightly and therefore inequality is decreasing. The policy-adjusted and emissions 3 scenarios are less favourable for income distribution. There are a number of reasons for this. The first reason is related to a higher cost of investment, the relative decrease in growth of other sectors in the economy has an impact on employment and ultimately household income. There is a negative impact on the growth of all sectors, except the electricity sector (as one would expect) and natural gas mining; driven by the increase in demand for gas turbines in the two alternative scenarios (policy-adjusted and emissions 3). Coal mining, for instance, contracts by 1.14 per cent relative to the base; as a sector with a high employment multiplier, especially for low-skilled labour, this would detract from the gains in the electricity sector. The second reason is directly linked to the decrease in employment of the various labour groups over the period. Renewable energy options are more labour intensive, per GWh of electricity, in comparison to base load coal, although they do require a larger proportion of high-skilled labour. There is a slight decrease in overall employment from the investment in the alternative plans, relative to the base case; the decrease in employment for the low-skilled labour group is much lower than the average. Low-skilled labour employment decreases by 5 per cent compared to the base, compared to high-skilled (individuals with tertiary-level education) that had no unemployment over the period under the emissions 3 case. This, in-turn, has a negative impact on income distribution.

The reduction in emissions, as one would expect, is significantly higher for the emissions 3 scenario, with a reduction of 18 per cent compared to the base[16]. As previously mentioned, at least R0.46 trillion would be required for the electricity sector to reach its emissions plateau by 2025, in addition to the R1.3 trillion already estimated for the emissions 3 scenario. The relatively high allocation of renewables in the policy-adjusted scenario does make a dent in South Africa's emissions, however, with a reduction of 11 per cent compared to the base.

Table 1.4: Simulation Results With a Carbon Tax (in percentage)

	GDP Growth	Inequality	Emissions Reduction	Employment
Base	3,90	1,06	-29,26	1,31
Policy	3,79	0,97	-39,66	1,30
Emissions 3	3,64	0,81	-43,62	1,28

Source: Own calculations

The simulation results with a carbon tax are shown in Table 4 and indicate that the tax is likely to have a slightly contractionary effect on the economy, with some sectors actually becoming more profitable given the changes in relative prices that occur as a result of the tax. Biomass, for example, grows by 2.38 per cent relative to the base for the base case scenario with a carbon tax; 2.05 per cent for the policy-adjusted and 1.47 per cent for the emissions 3 scenario.[17] Given that the effective tax rate is overestimated in these simulations, a conclusion can be made that the tax may not have a detrimental effect on the economy and could incentivise growth in 'cleaner' sectors; highlighting the potential benefit of moving to a low carbon trajectory.

The reduction in emissions is significantly increased for all three cases, with approximately a 44 per cent reduction in emissions in the emissions 3 scenario by 2030, relative to the base. The tax is also very effective in reducing emissions in the base case, with a reduction of 30 per cent. The results echo that found in previous studies, that even at the full R120 per ton of CO2 and with a very costly electricity build plan based on a carbon limit for the sector, South Africa is unlikely to reach their target of a 42 per cent reduction in emissions by 2025, relative to a 'business-as-usual' baseline. One can conclude that the proposed tax level, even without the 'basic tax-free threshold' and complex exemptions for energy-intensive users, is still too low and needs to be revised if South Africa wants to reach its emissions targets.

The distributional impact of a carbon tax is not as favourable; however, the income gap is still narrowing. Employment also remains positive, albeit less than the employment growth rate without a carbon tax. The slight decrease is attributed to the marginal contraction of the economy due to increased energy prices.

There are a number of tax recycling mechanisms that are available to increase the distributional impact of the carbon tax; referring back to Alton *et al.* (2012) where it was found that the revenue recycling option is an important driver of the economic impact of a carbon tax. Given the findings of their study one would expect that the distributional impact of the carbon tax would be more favourable if the revenue was recycled to fund social grants and less favourable if it were coupled with a decrease in corporate tax. A complete analysis of potential revenue recycling options has been noted for future work.

Conclusion

In conclusion, the introduction of renewable energy and low carbon trajectories is likely to have a slightly negative impact on employment and a marginally contractionary impact on the economy. This is a key finding as it indicates that the implementation of these mitigation actions is not likely to cripple the economy and that there are benefits that South Africa should capitalise on.

Renewable energy options, unfortunately, still have relatively high investment costs. These costs are the main driver for the results in this study. The higher cost of renewables causes a slightly contractionary effect on the economy from the decrease in the investment funds available to other more profitable sectors.

And regarding employment, even though some renewable energy options have higher job years per MW (approximately 52 job years/MW for PV compared to 10.8 for coal-fired plants), the positive impact on direct employment is drowned out by the negative impact on indirect employment. The loss of low-skilled jobs dominates this effect, which results in higher income inequality.

In terms of emissions reduction, one can conclude that the introduction of renewable energy, even to the extent proposed in the emissions three scenario, is not sufficient for South Africa to meet its emissions reduction target of 42 per cent against a 'business-as-usual' baseline by 2025.

The implementation of a carbon tax is likely to have less of a '*devastating*' impact than was previously thought. Higher energy prices might incentivise the development of 'cleaner' sectors such as the biomass industry. The addition of a carbon tax proves quite effective in terms of lowering total emissions; however, the tax level (even without the exemptions) is still too low and will not be enough to get emissions down to the target trajectory. Modelling a carbon tax of around R12 to R48 per ton of CO_2, the effective tax rate taking all proposed exemptions into account, would have even less of an impact on the emissions. The argument that an increased tax level will cripple the economy seems unjustified and South Africa should capitalise on the growth of sectors that could become profitable with the introduction of a carbon tax.

The distributional impact of a carbon tax is not favourable in this case, albeit the income gap is still narrowing and employment is still positive. Revenue recycling options are a key driver of impact of a carbon tax on the economy. Designing the carbon tax with a revenue recycling option to fund social grants is likely to lead to more favourable welfare effects, but less economic growth.

In conclusion, this paper shows that current renewable energy plans and the proposed carbon tax level are not enough to allow South Africa reach its emissions reduction target of 42 per cent by 2025. Both of these mitigation actions are found to have a less 'devastating' impact on the economy than was previously thought. If South Africa is to meet the challenge of decreasing emissions as well as decreasing inequality and eradicating poverty a higher carbon tax should be introduced along with a revenue recycling mechanism that could increase the income allocation to lower income deciles and result in increased welfare.

Notes

1. The model development initially started during a research stay at UNU-WIDER in Helsinki, where Dr James Thurlow and I disaggregated the energy sector in the e-SAGE model to include more detail for renewable energy generation technologies. Further model developments and the analysis of the potential implications of a carbon tax were conducted during my time at the German Institute of Global and Area Studies (GIGA) in Hamburg under the supervision of Jun.-Prof. Dr Jann Lay and funded by the Volkswagen Foundation.

2. World Bank (2013). "World Development Indicators 2013." Washington, D.C.: World Bank. http://data.worldbank.org. Accessed July, 2014.
3. Stats SA (2013). "Millennium Development Goals, Country Report 2013". Pretoria: Statistics South Africa. http://www.statssa.gov.za/MDG/MDGR_2013.pdf. Accessed July, 2014.
4. This increases to 34,6% if you include discouraged workers. Source: Stats SA (2015). "Quarterly Labour Force Survey: Quarter 4, 2014". Pretoria: Statistics South Africa. http://www.statssa.gov.za/publications/P0211/P02114thQuarter2014.pdf. Accessed July, 2014.
5. For further reading on the potential impacts of climate change in South Africa and Africa in general, see Gbetibouo and Hassan (2005) and Bryan et al. (2009).
6. Multiple-criteria decision making or multiple-criteria decision analysis (MCDA) refers to a method of structuring and solving decision and planning problems that involve multiple criteria. In this case, the IRP engaged with stakeholders to assign a score for each scenario in terms of the aforementioned criteria. The scores were weighted, aggregated and the scenarios compared according to their overall scores.
7. The economy-wide model used in this study, e-SAGE, is a computable general equilibrium model and is calibrated using actual economic data for the South African economy. CGE models are widely used for policy analysis. For further reading, see Thurlow (2004).
8. The partial-equilibrium model used in Pauw (2007) was a MARKAL model for South Africa's energy sector. The MARKAL model is a long-term multi-period energy technology optimization model. Selected results such as changes in the energy supply mix, energy efficiency and investment requirements from the energy model were used to inform the CGE model.
9. The IRP scenarios were modeled using PLEXOS Integrated Energy Model, a mathematical optimization model for the energy sector.
10. Gross operating surplus is the portion of income that is earned by the capital factor from production by incorporated enterprises.
11. These assumptions are based on the lifetime of the plant and are based on EPRI (2010) and, for renewable energy options, REIPPPP announcements (DoE, 2013).
12. Levelized cost of plant is the unit cost of electricity generation over the life of the plant. It includes all the costs needed to build and operate a power plant over its lifetime, normalized over the total net electricity generated by the plant.
13. The portion of investment assumed to flow out the economy through imported content requirements during the build phase. Based on weighted averages for imported content over the first 2 bids (DoE, 2013).
14. The IRP has recently been criticized for being 'out-of-date', especially in terms of the demand forecasts and the cost assumptions for the technology options; the Renewable Energy IPP Procurement Programme provides more realistic employment, local content and cost data. The estimates given in the table will be updated to reflect these in the near future.
15. The use of this form of inequality measure may be criticized for being over-simplified and vulnerable to the effects of outliers. For the purpose of this paper it is sufficient and more complex inequality measures could be used in future modeling exercises.

16. These are economy-wide emissions, not only for the electricity sector.
17. This finding is supported by other research and existing policy that identify the potential for growth in South Africa's biomass sector; for further reading, see Winkler (2005), Dasappa (2011) and DME (2007).

References

Alton, T., Arndt, C. Davies, R. Hartley, F., Makrelov, K., Thurlow, J., Ubogu, D., 2012, 'The Economic Implications of Introducing Carbon Taxes in South Africa', *UNU-WIDER Working Paper* No. 2012/46. United Nations University's World Intitute for Development Economics Research (UNU-WIDER), May 2012.

Arndt, C., Davies, R., & Thurlow, J., 2011, *Energy Extension to the South Africa General Equilibrium (SAGE) Model*. United Nations University's World Intitute for Development Economics Research (UNU-WIDER).

Armington, P., 1969, *A Theory of Demand for Products Distinguished by Place of Production*, IMF Staff Papers, Vol. 16, No. 1,, pp. 159:178.

Bryan, E., Deressa, T.T., Gbetibouo, G.A., Ringler, C., 2009, 'Adaptation to Climate Change in Ethiopia and South Africa: Options and Constraints', *Environmental Science & Policy*, Vol. 12, No. 4, pp. 413-426.

Dasappa, S., 2011, 'Potential of Biomass Energy for Electricity Generation in Sub-Saharan Africa', *Energy for Sustainable Development*, Vol. 15, No. 3, pp. 203-213.

Dervis, K., de Melo, J., & Robinson, S., 1982,. *General Equilibrium Models for Development Policy*, Washington, D.C.: A World Bank Research Publication.

Devarajan, S., D. S. Go, S. Robinson, and K. Thierfelder, 2011, 'Tax Policy to Reduce Carbon Emissions in a Distorted Economy: Illustrations from a South Africa CGE Model', *The B. E. Journal of Economic Analysis and Policy*, Vol. 11, No. 1, pp. 1-22.

DoE, 2010, *Integrated Resource Plan for Electricity*, Version 8 Draft Report, 8 October,

DoE, 2011, *Integrated Resource Plan for Electricity*. Final Draft.

DoE, 2013, *Renewable Energy IPP Procurement Programme;* Bid Window 3 Preferred Bidder's Announcement. Department of Energy, 4 November:

DME, 1998, *White Paper on the Energy Policy of the Republic of South Africa, Pretoria:* Department of Minerals and Energy.

DME, 2007, *Biofuels Industrial Strategy of the Republic of South Africa, Pretoria:* Department of Minerals and Energy.

EPRI, 2010, *Power Generation Technology Data for Integrated Resource Plan of South Africa, Pretoria:* Electric Power Research Institute.

Eskom, 2011, *Annual Report*. Eskom.

Gbetibouo, G.A., Hassan, R.M., 2005, 'Measuring the economic Impact of Climate Change on major South African Field Crops: A Ricardian Approach', *Global and Planetary Change*, Vol. 47, Vols. 2-4, pp. 143-152.

Kearney, M., 2010, *Modelling the Impact of CO_2 taxes in Combination with the Long Term Mitigation Scenaios on Emissions in South Africa Using a Dynamic Computable General Equilibrium Model*. Conference Proceedings on Putting a Price on Carbon, Cape Town: Energy Research Centre, UCT.

Lofgren, H., Harris, R. L., & Robinson, S., 2001, *A Standard Computable General Equilibrium (CGE) Model.*

Nell, K., 2003, 'Long-run Exogeneity Between Saving and Investment: Evidence from South Africa', *Working Paper 2-2003 on Trade and Industrial Policy Strategies,* Johannesburg, South Africa.

Pauw, K., 2007, *An Input into the Long Term Mitigation Scenarios Process – LTMS Input Report 4,* Cape Town: ENERGY RESEARCH CENTRE, University of Cape Town.

Republic of South Africa (RSA), 2010, *Reducing Greenhouse Gas Emissions: The Carbon Tax Option,* Pretoria: National Treasury, Government of the Republic of South Africa.

Republic of South Africa (RSA), 2013, *Carbon Tax Policy Paper: Reducing Greenhouse Gas Emissions and Facilitating the Transition to a Green Economy* – Policy Paper for Public Comment, Pretoria: National Treasury, Government of the Republic of South Africa, May.

Thurlow, J., & van Seventer, D. E., September 2002, *A Standard Computable General Equilibrium Model For South Africa,* Washington, D.C.: Trade and Macroeconomics Division, International Food Policy Research Institute.

Thurlow, J., 2004, *A Dynamic Computable General Equilibrium (CGE) Model for South Africa: Extending the Static IFPRI Model,* Pretoria: Trade and Industrial Policy Strategies (TIPS).

Thurlow, J., 2008, *A Recursive Dynamic CGE Model and Microsimulation Poverty Modue for South Africa.* Washington, D.C: International Food Policy Research Institute.

Winkler, H., 2005, 'Renewable Energy Policy in South Africa: Policy Options for Renewable Electricity', *Energy Policy,* Vol. 33, No. 1, pp. 27-38.

2

Les migrants climatiques en quête d'adaptation : les éleveurs Mbororo émmigrent en Rd Congo

Félicien Kabamba Mbambu

Introduction

Une grande partie de l'agriculture africaine est une agriculture sous pluie. Avec l'accélération des changements climatiques, de grandes perturbations dans l'articulation des saisons de culture frappent durement la sécurité alimentaire. Si la production agricole nécessaire à l'alimentation humaine est de plus en plus affectée, celle destinée à l'alimentation animale est devenue à la fois rare et coûteuse. Cette carence pose de réels problèmes d'adaptation aux éleveurs africains habitant des régions qui subissent de plein fouet les méfaits des changements climatiques, changements qui ont induit des modifications des conditions de température, de régime des saisons culturales, ainsi que de l'organisation pastorale dans son ensemble.

Poussés par l'austérité des changements climatiques, ces pasteurs émigrent vers la partie centrale du continent africain. Arrivés en République démocratique du Congo vers l'an 2000[1], ils habitent le nord-est du pays, dans la région de l'Uélé située dans la Province orientale, une zone remarquable par ses savanes boisées caractérisées par l'articulation naturelle des savanes et de la forêt.

Souvent définis à tort comme peuple nomade, ces pasteurs Mbororo se sont nettement sédentarisés en République démocratique du Congo. Ils habitent la région de l'Uélé de façon permanente. On n'observe plus de mouvements migratoires vers d'autres territoires. Du coup, le nomadisme apparaît comme une notion conjoncturelle commandée par des facteurs géophysiques et climatiques.

Le postulat de base qui guide notre réflexion repose sur le fait que l'activité pastorale est un déterminant socioéconomique de base pour ces populations, et que, devant la vulnérabilité due aux changements climatiques, la migration s'impose comme mécanisme d'adaptation.

Cette migration engendre plusieurs problèmes : i) Conflits fonciers récurrents entre agriculteurs et éleveurs, ii) conflits entre les deux modes de production très antagonistes, iii) problèmes de citoyenneté étant donné que le Mbororo ne se définit ni comme Congolais, ni comme étranger, mais plutôt, comme citoyen du monde à la recherche du pâturage pour son bétail, iv) défis sécuritaires dans une région qui était déjà en proie aux conflits armés il y a plus de deux décennies.

La République démocratique du Congo, issue d'un conflit armé de plus de 20 ans, se prépare à un autre type de conflit, celui provenant des changements climatiques.

Ce papier se propose de tracer un cheminement réflexif du phénomène migratoire Mbororo dans sa dimension à la fois socioéconomique et sécuritaire. Sans occulter les enjeux identitaires et fonciers, l'analyse qui en résulte s'éloigne des lectures partielles et sélectives qui considèrent les migrations africaines uniquement selon l'axe sud-nord. Et pourtant, les constatations de l'histoire récente débouchent également sur une nouvelle reconfiguration du phénomène migratoire dans l'hémisphère sud en raison de la quête permanente des conditions climatiques adaptées aux modes de production en vigueur dans la région.

Recadrage du discours sur la migration environnementale

Le phénomène migratoire a été au cœur de plusieurs travaux qui ont donné lieu à de nombreux courants paradigmatiques au cours des quatre dernières décennies. Le foisonnement de cadres théoriques aussi différents alimente les réflexions sur un phénomène présenté comme mondial, mais qui a toujours une explication liée au contexte local, donnant ainsi lieu à une diversité des cadres référentiels.

La revue de ces théories tant du point de vue de l'évolution des approches que de l'apparition des nouvelles formes des migrations démontre le caractère à la fois dynamique et prolifique du chantier théorique.

La référence aux paradigmes micro-individuels et macrostructurels

Au cours de la décennie 1960, les premiers travaux sur les migrations s'inscrivaient dans une approche micro-individuelle et mettaient en perspective la décision individuelle :

> Avant de prendre la décision de quitter son lieu de résidence, l'individu examine les coûts et les bénéfices liés à la migration potentielle. Cette approche est souvent associée au texte de Larry Sjaastad publié en 1962, dans lequel il se propose d'identifier les coûts et les bénéfices et de déterminer le « retour sur investissement » résultant des migrations. (Victor Piché 2012:154)

Ces travaux, qui tentaient de mettre en exergue les motivations individuelles, ont fait place à la théorie de la transition démographique, qui s'est longtemps efforcée de forger une explication du phénomène migratoire par les changements dans les niveaux de fécondité et de mortalité.

Les migrations sont alors envisagées sous l'angle des coûts et des bénéfices par rapport au migrant pris individuellement. Il s'ensuit que pareille conception des migrations est très réductionniste car elle réduit l'explication du phénomène migratoire à un niveau minimaliste.

Aujourd'hui, le phénomène de la mondialisation a permis d'infléchir l'analyse vers des facteurs plus globalisant à l'instar des évolutions qui affectent le secteur de transport ainsi que les nouvelles technologies de l'information.

C'est dans l'exploration des nouveaux paradigmes plus ouverts que voit le jour l'approche macrostructurelle. Le schéma analytique proposé « tente d'identifier tous les éléments de l'environnement pouvant affecter les mouvements migratoires, allant de l'environnement économique à la technologie, à l'environnement social et enfin aux facteurs politiques » (Victor Piché 2012:157). La prise en compte de la variable circulation de l'information va beaucoup influencer les nombreux travaux sur les réseaux sociaux et familiaux comme déterminants des phénomènes migratoires.

La valorisation de l'approche socioenvironnementale

Dans la perspective d'une exploration plus complète des facteurs des migrations, les changements climatiques et leurs effets sur les dynamiques migratoires ont conduit à réorienter la perception des phénomènes migratoires :

> Ils replacent la lutte contre le changement climatique, la désertification et la dégradation des terres au cœur de la crise globale de l'écosystème et montrent l'interdépendance des processus de dégradation des terres, des eaux et de la biomasse, du changement climatique, de la sécurité alimentaire et de la lutte contre la pauvreté. (Grigori Lazarev 2009:2)

En établissant un lien entre migration et développement, le concept de stratégie de subsistance a été forgé :

> Une stratégie de subsistance peut alors être considérée comme le choix, stratégique ou délibéré, effectué par un ménage ou par ses membres individuels, de recourir à une série d'activités pour préserver, sécuriser et améliorer ses/leurs moyens de subsistance. Ce choix particulier se base sur l'accès (sélectif) aux actifs, la perception du champ des possibilités et les aspirations des acteurs. Comme ces aspects varient selon les ménages et les individus, les stratégies de subsistance peuvent paraître particulièrement hétérogènes. (Haas 2010:22)

Dans ce contexte, un nombre croissant de travaux a défini la migration comme une composante majeure des stratégies utilisées par les ménages pour diversifier, sécuriser et, éventuellement, améliorer durablement leurs moyens de subsistance.

La migration est souvent associée à d'autres stratégies, telles que l'intensification agricole et les activités non agricoles locales.

Perçue comme ingénierie sociale reposant largement sur des mécanismes permettant de faire face aux vulnérabilités dues à la dégradation des territoires, la migration est saisie comme une stratégie d'adaptation aux changements climatiques dans un contexte où la lutte contre la désertification et la dégradation des terres est au cœur de la crise globale de l'écosystème.

> L'Organisation internationale des migrations […] définit les « migrants environnementaux » comme correspondant « aux personnes qui, pour des raisons de changement soudain ou progressif de l'environnement qui affectent leurs vies ou leurs conditions de vie, sont forcées de quitter leurs habitations, de façon temporaire ou permanente, et se déplacent dans leur pays ou à l'extérieur ». Les experts avancent un chiffre de 200 millions de migrants environnementaux au cours du siècle. (Gregori Lazarev 2009:38)

En effet, les travaux empiriques indiquent que la migration est souvent un choix délibéré « visant à améliorer les moyens de subsistance » (Bebbington 1999:2 027) et à « atténuer les fluctuations du revenu familial, souvent sujet aux caprices climatiques » (De Haas et al. 2000:8 cité dans Haas 2010:23).

La migration peut ainsi être conçue comme un moyen d'acquérir une plus grande variété d'actifs, qui jouent le rôle de garantie contre les chocs et les difficultés futurs.

D'autre part, les impacts humains des dérèglements climatiques et les inégalités qu'ils induisent sont encore insuffisamment étudiés. Les Nations unies ont joué un rôle de pionnier dans l'éveil des consciences sur les changements climatiques. Les terres arides représentent 36 pour cent du stock total de carbone des écosystèmes terrestres.

> L'adoption rapide de la Stratégie décennale de l'UNCCD, lors de la 8e Conférence des parties à Madrid en septembre 2007, a créé une nouvelle dynamique pour prendre en compte leur importance. Elle propose en effet à toutes les parties prenantes de la Convention une plateforme revitalisée d'action commune, en particulier en mettant fortement l'accent sur la réduction de la vulnérabilité des personnes et des écosystèmes, et, en même temps, en mettant en évidence les effets positifs globaux de la lutte contre la DLDD. La restauration de la fertilité des sols et la conservation de l'eau sont en effet des conditions sine qua non, d'une amélioration de la productivité des terres arables, elles constituent la base de toute la chaîne dont dépend la sécurité alimentaire. (Grigori Lazarev 2009:4)

Émigrer pour s'adapter

Au cours des trente dernières années, les pays du Sahel ont connu des sécheresses et des crises alimentaires sévères qui ont perturbé toute la filière agricole et pastorale. L'absence de pluie et la pénurie de récoltes ont mis tout le Sahel dans une situation d'insécurité alimentaire dont les effets se manifestent diversement selon les secteurs d'activité. L'allongement de la durée de la saison sèche combiné au déficit d'encadrement en matière de pastoralisme et aux politiques irréalistes en matière d'accès à l'eau a provoqué une véritable mutation dans le mode de vie des populations en général et des éleveurs en particulier. Plus récemment, les années 2009, 2011 et 2013 ont été marquées par un allongement très accentué de la durée de la saison sèche.

Les changements les plus couramment observés sont la variation de l'intensité, de la fréquence et de la durée des précipitations, la diminution de la pluviométrie, l'assèchement des pâturages et des réserves d'eau.

> La rapidité avec laquelle se manifestent les changements climatiques impose l'option de l'adaptation comme celle du réalisme et de l'évidence. (Tubiana, Gemenne & Magnan 2010:1)

Étant donné l'incertitude de l'évolution du climat, émigrer vers le centre du continent africain est une forme d'adaptation aux changements climatiques qui deviennent, pour les éleveurs, le plus grand défi de l'histoire. Cela constitue à la fois un mécanisme de survie et une anticipation par rapport aux perturbations vécues et futures qui augmentent de façon permanente la vulnérabilité des populations et des activités auxquelles elles s'adonnent. On constate cependant une modification des parcours de transhumance qui laisse penser à une option écologique inévitable. C'est ainsi que sont observées d'importantes vagues de migrations des éleveurs Mbororo vers la partie équatoriale du continent africain, en particulier dans la région de l'Uélé en République démocratique du Congo où la présence d'abondants pâturages constitue une attraction et un espoir vital.

En revanche, si les migrations climatiques sont présentées comme une panacée, les conflits qui en résultent font du changement climatique un facteur d'insécurité, de tensions et de conflits permanents. Envisagée dans une perspective d'adaptation au changement climatique, l'insécurité sociopolitique est une circonstance aggravante du contexte de vulnérabilité. Non seulement elle se pose comme contrainte à l'adaptation naturelle, mais elle oppose au processus décisionnel des limites conjoncturelles.

Migrations pastorales, épreuves locales et contraintes juridiques

Les Mbororo qui ont immigré dans le nord-est de la République démocratique du Congo viennent de plusieurs pays de la région et principalement de la République centrafricaine, du Tchad et du Soudan. Dans ces trois pays, l'activité pastorale est largement répandue et représente une richesse importante qui contribue significativement au revenu national : « le Tchad disposerait d'un cheptel bovin estimé à plus de 20 millions de têtes de bétail, il représenterait ainsi entre 15 et 20 % du produit intérieur brut » (International crisis group 2014:3). Si les pasteurs tchadiens étaient déjà connus comme éleveurs nomades, la sécheresse et les changements climatiques qui ont durement affecté ce pays, avec comme corollaire l'assèchement de plus de 3/4 de la superficie du lac Tchad, ont accéléré leurs migrations vers le sud dans l'axe RCA-RDC[2].

Ces déplacements massifs des hommes et des troupeaux entraînent des tensions et des conflits qui augmentent le climat d'insécurité dans une région en proie à une instabilité croissante depuis les années 1990. En effet : « les pasteurs ont payé un lourd tribut aux guerres civiles qui ont frappé le Tchad dans les années 1980 et le Darfour au début des années 2000 : les éleveurs étaient souvent rackettés par les rebelles ou, au contraire, privés de leurs troupeaux par les forces armées au nom de « l'effort de guerre » (International crisis group 2014:4).

Cette transhumance occasionne également des dégâts écologiques ainsi que la compétition pour l'accès à la terre qui dégénère souvent en affrontements armés avec les communautés locales.

Une immigration au forceps

En République démocratique du Congo, les Mbororo sont considérés comme des envahisseurs. Leur première tentative d'émigrer en République démocratique du Congo date des années 1980. Ils avaient alors été chassés par l'armée congolaise. Profitant des conflits armés avec l'occupation de la partie nord du pays par les rebelles du Mouvement de libération du Congo et du Rassemblement congolais pour la démocratie, ils vont par vagues successives, occuper le nord-est du pays au cours de la décennie 2000. Ils s'infiltreront progressivement dans la région de l'Uélé par les pistes frontalières de Obo en RCA vers Passi en RDC, et de Mboki en RCA, vers la rivière Gwane en RDC, après avoir traversé la rivière Mbomu.

À la recherche des pâturages

En quête des meilleurs pâturages pour leur bétail, ils se sont orientés vers deux axes.

D'abord, le district des Bas-Uélé où ils ont choisi de s'implanter dans les territoires d'Ango et Poko. Le territoire d'Ango est le plus vaste de ce district avec une superficie de 34 704 km[2] et une population de plus de 70 000 habitants pour une densité de +50 habitants par km[2]. C'est une zone du pays qui est en partie savanicole alors qu'une autre bonne partie est forestière. Elle présente des avantages certains pour l'activité tant agricole que pastorale. Les populations habitant cette partie du territoire sont des agriculteurs qui s'adonnent à la culture de riz, d'haricots et de manioc. Le pastoralisme comme mode de vie est absent des pratiques culturales locales.

Le deuxième axe est le district de Haut-Uélé et là, ils se sont orientés précisément vers les territoires de Dungu et de Faradgé. Cette région d'Uélé possède une hydrographie attirante pour l'activité pastorale. Elle est composée de la rivière Bamokandi au sud, la rivière Mbamu au nord et de la rivière Uélé à l'Est. L'ensemble du bassin hydrographique est très dense avec ses ramifications en cours d'eaux qui couvrent une bonne partie de la région. Cette région connaît huit mois de saison de pluie, soit d'avril à novembre, et quatre mois de saison sèche, de décembre à mars :

> La végétation est constituée de forêts qui alternent sous forme de mosaïque avec des espaces considérables couverts de hautes herbes. La forme caractéristique est la savane arborée. Le tapis herbacé est formé de graminées pouvant atteindre 2 à 3 mètres de haut et dont la densité peut être élevée ou faible. C'est le domaine des mammifères herbivores et des grands carnassiers. C'est aussi un domaine potentiel d'élevage bovin. Le tapis herbeux est dominé par une strate arbustive faite d'arbustes plus ou moins espacés, perdant leurs feuilles en saison sèche, mais résistante aux incendies. (Nkoy Elela 2007:22)

Cette partie du pays est majoritairement habitée par l'ethnie Zandé qui est ethnologiquement assimilée au groupe soudanais. C'est donc l'existence de grandes étendues de pâturages et d'une hydrographie naturellement abondante qui constitue le principal motif d'attraction des Mbororo vers cette région et par dérivation l'explication la plus rationnelle du mouvement migratoire des Mbororo dans les hauts et les bas Uélé.

Le contexte légal et réglementaire de l'immigration pastorale

Selon la législation congolaise, immigrer en RDC procède d'une démarche individuelle marquée par l'examen minutieux des cas souvent pris individuellement. Et dans cette optique, toute immigration collective ne peut être envisagée que dans des cas très limités[3]. Or, la particularité du mouvement migratoire Mbororo est qu'il est à la fois collectif et violent. Il concerne des familles entières et leurs dépendances. Il concerne aussi des milliers de têtes de bétail en quête de pâturages. Ces migrants sont souvent armés ou finissent par le devenir en échangeant du bétail contre les armes dans une région où le trafic d'armement est assez courant. La porosité des frontières nationales leur permet de pénétrer le territoire national et par la suite de solliciter des autorisations auprès des autorités locales.

La RD Congo ne dispose pas de code pastoral. Ce sont donc les dispositions relatives au code agricole qui s'appliquent au secteur de l'élevage. D'ailleurs, le secteur agricole et pastoral n'avait pas fait l'objet d'une législation sectorielle spécifique jusqu'en décembre 2011 où un code agricole a été promulgué. Les titres agricoles étaient, avant l'adoption de ce texte, attribués conformément aux dispositions de la loi foncière, comme des titres fonciers. L'emphytéose, d'une durée de 25 ans renouvelable, avait été jusqu'à cette date le titre qui donnait accès à l'exploitation des terres à des fins agricoles ou pastorales.

À ce jour, et en vertu du nouveau code agricole, une nouvelle catégorie de titre agricole a vu le jour, à savoir : la concession agricole. La concession agricole est reconnue et définie par la nouvelle loi n° 11/022 du 24 décembre 2011 portant principes fondamentaux relatifs à l'agriculture comme *un contrat ou convention conclu(e) entre l'État et un opérateur agricole, permettant à ce dernier d'exploiter le domaine privé de l'État dans des limites précises, en vue d'assurer la production agricole* (République démocratique du Congo 2011, article 6). Il existe trois sortes de concessions agricoles : i) les concessions agricoles d'exploitation industrielle, ii) les concessions agricoles d'exploitation de type familial et iii) les concessions agricoles d'exploitation familiale. Les concessions agricoles de type industriel impliquent une étendue assez vaste, les moyens en hommes et en matériels visant un important potentiel de production. Les concessions agricoles de type familial consistent dans une exploitation agricole familiale faisant appel à une main-d'œuvre contractuelle et constituant une unité de production d'une capacité moyenne. Les concessions agricoles familiales consistent dans une exploitation agricole dont le personnel est constitué des membres de la famille de l'exploitant (République démocratique du Congo 2011, article 14). La superficie maximale des concessions d'exploitation agricole industrielle reste régie par les dispositions de la loi foncière, étant donné le silence de la loi agricole. Celles des concessions agricoles familiales et de type familial seront déterminées par arrêté du gouverneur de province, en tenant compte des particularités de la province.

Par ailleurs, faute d'une disposition explicite de la loi prévoyant l'adjudication, la procédure de gré à gré reste le seul mode d'acquisition des terres à des fins agricoles et pastorales, quelle que soit la catégorie de concession agricole ou pastorale. La loi n° 11/022 susmentionnée, en son article 16, prévoit que les concessions agricoles sont concédées aux exploitants et mises en valeur dans les conditions prévues par la loi. Mais elle ne précise pas de quelle loi il s'agit.

Par ailleurs, ce cadre formel et institutionnel d'administration des terres, déjà complexe, entre en compétition avec un autre système de gouvernance foncière, relevant cette fois-ci des coutumes et traditions locales, et comportant des usages des terres d'ordre clanique et familial, répondant à des besoins locaux.

Dans le contexte particulier des migrations Mbororo, aucune négociation entre les communautés locales et les migrants Mbororo n'a été entamée pour permettre aux Mbororo d'accéder paisiblement aux terres traditionnellement reconnues par les lois nationales comme terres coutumières.

Cette exigence de négociation avec les communautés est d'ailleurs le fondement des droits foncier coutumier en RD Congo, car, après avoir supprimé la catégorisation des terres retenue par les textes coloniaux[4], la loi du 20 juillet 1973 en avait introduit une autre, faite d'une part des « terres urbaines » et, de l'autre, des « terres rurales ». Les droits fonciers des communautés locales s'exercent en principe sur les terres rurales.

Ce sont, en effet, les dispositions des articles 387 à 389 de la loi du 20 juillet qui règlent le sort des droits que les communautés locales détiennent désormais sur les terres qu'elles occupent. Ils sont respectivement libellés comme suit :

- « Les terres occupées par les communautés locales deviennent, à partir de l'entrée en vigueur de la présente loi, des terres domaniales » (Loi foncière 1973, article 387) ;
- « Les terres occupées par les communautés locales sont celles que ces communautés habitent, cultivent ou exploitent d'une manière quelconque – individuelle ou collective – conformément aux coutumes et usages locaux » (Loi foncière 1973, article 388) ;
- « Les droits de jouissance régulièrement acquis sur ces terres seront réglés par une ordonnance du président de la République » (Loi foncière 1973, article 389).

La domanialisation des terres par l'État congolais opérée en 1967 n'a pas eu pour conséquence d'abolir les droits fonciers des communautés locales. Elle en a plutôt changé la teneur : au lieu du droit d'occupation sur les terres dites indigènes, les communautés locales exercent désormais un droit de jouissance collectif sur les terres du domaine privé de d'État (Mpoyi Mbunga 2012:17).

Les coutumes et usages locaux constituent le socle de la vie au village et les conflits se règlent en application de la coutume ; de sorte que l'on est fondé à affirmer que les droits fonciers des communautés locales sur les terres qu'elles occupent sont des droits coutumiers, alors même qu'ils trouveraient leurs premiers fondements dans la loi et dans la constitution[5].

Si les migrations pastorales des Mbororo sont illégales, la RD Congo ne dispose pas encore de code pastoral censé régler les situations de transhumance étant donné que la plupart des éleveurs congolais sont plutôt sédentaires. L'apparition de ce phénomène dans le nord du pays suggère donc de couvrir juridiquement des situations jusque-là absentes du cadre normatif congolais. Si ailleurs, comme au Tchad, les couloirs de transhumance ont été parfois créés pour limiter les dégâts des migrations pastorales, aucune mesure d'encadrement n'est encore envisagée, soit dans le cadre de la législation nationale, soit encore dans l'encadrement réglementaire de la transhumance.

Détérioration des relations entre Mbororo et communautés locales

Les Mbororo sont présentés comme un peuple éleveur. Cependant, une telle généralisation est loin de refléter la composition hétérogène de ce groupe. Parmi eux, on retrouve des braconniers, des anciens rebelles, des commerçants, des trafiquants d'armes, etc.

Leur mode de vie est localement jugé incompatible avec la culture locale. Une différence de civilisation apparaît et affecte durement les relations avec les populations locales.

Une difficile intégration dans la société congolaise

La cohabitation entre Mbororo et communautés locales est quasiment conflictuelle. Ces premiers disposent des armes et en usent souvent, soit pour forcer l'accès aux terres, soit pour imposer leur mode de production pastoral dans les nouveaux espaces marqués par la pratique de l'agriculture familiale.

« Localement, les Mbororo sont jugés à la fois agressifs et non respectueux des normes et valeurs traditionnelles locales ». (Nkoy 2014:44) Le récit de Désiré Nkoy sur leurs modes de vie témoigne d'une difficile intégration due à l'existence des modes de production très antagonistes (Nkoy 2014:44) :

> i) Ils détruisent tous les pièges, tendus par la population, qu'ils rencontrent en brousse, prétextant que ces pièges pourraient attraper et tuer leurs vaches ; ii) Ils chassent les femmes qui vont en brousse pour faire la pêche à la digue, car disent-ils, elles assèchent les eaux que devraient boire les vaches ; iii) Ils détruisent toutes les ruches des abeilles d'où la population recueille du miel, parce que le miel est un poison pour les vaches ; iv) Ils tuent tous les herbivores qu'ils rencontrent en brousse pour préserver le pâturage en faveur de leurs vaches. (loc. cit.)

De culture polygamique, ce sont des peuples à famille nombreuse. Une seule famille peut compter jusqu'à 50 membres. Une fois installés, il leur est reproché d'accueillir illégalement les nouveaux venus pour faciliter leur accès aux terres. Cependant, ces arrivées incontrôlées et massives ont créé un certain déséquilibre démographique dans une région où la densité de la population est souvent faible. Les Mbororo ont donc constitué dans cette contrée, une catégorie distincte du reste des communautés locales. On ne trouve pas encore de collaboration et de complémentarité socioculturelles entre éleveurs et agriculteurs ou encore des liens de mariage avec les communautés locales.

Rupture de l'équilibre démographique

Le Bas-Uélé et le Haut-Uélé sont les deux districts qu'ils ont occupés. Le Haut-Uélé a une densité assez élevée environ 21 habitants/Km² alors que le Bas-Uélé est relativement moins peuplé avec 8 habitants/Km². Les tensions les plus fréquentes sont plutôt rapportées dans le district de Haut-Uélé.

Tableau 2.1 : Démographie dans les territoires des Bas-Uélé et Haut-Uélé

Entités administratives	Hommes	Femmes	Populations totales
District du Bas-Uélé	436 877	483 170	920 0477
1. Territoire de Buta	59 954	65 474	125 428
2. Territoire d'Aketi	75 456	82 404	157 860
3. Territoire de Bongo	94 119	102 782	196 901
4. Territoire de Ango	36 412	35 840	71 892
5. Territoire de Bambesa	80 293	97 683	167 976
6. Territoire de Poko	90 643	98 987	189 630

District de Haut-Uélé	725 816	783 157	1 508 973
1. Territoire de Rungu	175 278	189 125	364 403
2. Territoire de Niangara	48 872	52 733	101 605
3. Territoire de Dungu	99 550	107 415	206 965
4. Territoire de Faradge	128 614	138 775	267 389
5. Territoire de Watsa	88 803	95 819	184 622
6. Territoire de Wamba	184 699	199 290	383 989

Source : Désiré NKOY

Le lien entre émergence des tensions et densité démographique traduit l'idée qu'avec la densité élevée, la raréfaction des ressources naturelles accroît la compétition pour leur contrôle. Ces tensions ont comme causes : l'accès à la terre, l'approvisionnement en eau pour le bétail, la difficulté de faire la chasse consécutivement à la présence d'importants cheptels dans l'aire de la chasse qui provoque l'éloignement des animaux sauvages, etc.

Tableau 2.2 : Effectifs des Mbororo en 2007

N°	District	Territoire	Effectifs
1.	Haut-Uélé	Dungu	2 500
		Faradgé	1 200
2.	Bas-Uélé	Ango	7 200
		Poko	6 750
		Total	16 500

Source : Désiré Nkoy

Le tableau ci-dessus indique que les Mbororo sont beaucoup plus nombreux dans les territoires de Ango et de Poko qui sont les moins peuplés alors que leur nombre dans les territoires de Dungu et de Faradgé est nettement inférieur. Les tensions les plus violentes sont fréquentes dans le territoire de Dungu où la densité est l'une des plus élevée (+ de 25 habitants/Km²). Par contre une tendance à la normalisation de la cohabitation est observée dans le territoire de Ango, territoire moins peuplé où on note pourtant une arrivée massive des troupeaux.

Violences, nomadisme et citoyenneté

La République démocratique du Congo a une superficie de 2 345 000 km². Elle partage plus de 10 500 km de frontières poreuses avec neuf pays voisins que sont la République du Congo, l'Angola, la Zambie, la Tanzanie, le Burundi, le Rwanda, l'Ouganda, le Soudan et la République centrafricaine. La plupart de ces pays ont connu des épisodes récurrents d'insécurité durant les 30 dernières années. Celles-ci ont affecté durement la République démocratique du Congo.

Une transhumance dans une région à faible présence étatique

Dans les territoires de Dungu et d'Ango qui sont frontaliers avec la République centrafricaine, la présence de l'État est minimale. Ainsi donc, l'organisation d'un pastoralisme emprunt de violence et de criminalité a trouvé dans cet espace sans État le lieu de prédilection pour l'implantation durable d'une activité qui se déploie dans un espace où l'érosion de l'appareil de l'État a laissé la place à l'émergence de la loi du plus fort.

Ainsi, la recrudescence de la criminalité pastoraliste est liée à l'effondrement des systèmes autoritaires. À propos de la déstabilisation des régimes à parti unique Paul Ango Ela pense que :

> Avec l'amorce du processus de démocratisation des années quatre-vingt-dix, l'orientation des menaces a fondamentalement changé dans la région. Le processus a entraîné des changements sociopolitiques qui ont ébranlé les régimes de parti unique. Il en a résulté la délitescence des « États mous » n'étant guère préparés à gérer des crises. (Ela Ango 1996)

La présence minimale de l'État dans cette partie du pays ne lui permet pas d'assurer la fonction de maintien de l'ordre et de la sécurité. Les mouvements rebelles qui avaient occupé cette région autrefois avaient juste maintenu un semblant d'ordre de manière à leur permettre l'exploitation économique du territoire. La corrélation entre l'émergence de la transhumance et la disparition progressive des appareils étatiques mérite d'être reconsidérée de ce point de vue.

Les violences liées à la transhumance ou la préfiguration d'une nouvelle génération des conflits en Afrique centrale.

La République démocratique du Congo est en proie à des conflits armés violents depuis 1996. Les nombreux observateurs estiment qu'ils trouvent leur origine dans la lutte pour le contrôle des ressources naturelles. Ce lien entre exploitation des ressources naturelles et conflits armés est plusieurs fois évoqué pour expliquer la prolifération des mouvements armés dans une vaste zone où l'État, en déliquescence, a cédé le terrain aux seigneurs de la guerre et aux milices armées dont le nombre et la dynamique opérationnelle sont devenus sujets à controverse. Parfois appelés « Première Guerre mondiale africaine » à cause de l'implication directe de plusieurs pays, les conflits à l'est de la RDC connaissent aussi l'immixtion d'un nombre très élevé de milices provenant des pays voisins. C'est les cas des Forces démocratiques de libération du Rwanda (FDLR), de l'Armée de résistance du Seigneur (LRA Ouganda) et des Forces démocratiques alliées/Armée nationale pour la libération de l'Ouganda (ADF/NALU). À côté de ces groupes armés prolifère un nombre élevé de petits groupes armés[6] organisés en force d'autodéfense et en milices armées.

Les conflits qui apparaissent avec le phénomène Mbororo sont plutôt d'une autre nature et ont des causes bien extérieures à la RD Congo dans la mesure où les éleveurs qui ont perdu leurs moyens de subsistance sont contraints à l'exode et donc à la quête des pâturages.

Une fois en RD Congo, ils négocient avec les seigneurs de la guerre. Pendant la rébellion du Mouvement pour la libération du Congo de Jean-Pierre Bemba, ils avaient réussi à négocier leur entrée sur le territoire national. Le MLC fut alors accusé de complicité. Des connivences ont été également constatées avec le mouvement rebelle ougandais LRA qui contrôlait un territoire au nord du pays.

Le lien entre migrations Mbororo et insécurité dans la région de l'Uélé provient aussi de l'inexistence d'un code pastoral occasionnant l'absence de balisage des couloirs de transhumance qui pouvaient pourtant permettre d'éviter l'empiétement des animaux sur les cultures familiales. Des nombreux champs sont, faute de cette précaution, régulièrement dévastés sans aucune contrepartie.

Nomade ou sédentaire ?

Les Mbororo sont-ils naturellement un peuple nomade ? Des constatations récentes amènent à remettre en cause le fondement d'un tel théorème. Le postulat de base sur lequel repose notre réflexion est qu'il y a une corrélation entre vulnérabilité climatique et transhumance.

Souvent définis à tort comme peuple nomade, les Mbororo se sont nettement sédentarisés en République démocratique du Congo. Ils habitent les territoires d'Ango, de Faradge, de Dungu et de Poko de façon permanente. On n'observe plus de mouvements migratoires vers d'autres territoires. L'abondance des pâturages et de grandes quantités d'eau dans cette région ne semble pas nécessiter des nouveaux épisodes de migrations. Du coup, le nomadisme apparaît comme une notion temporelle commandée par des facteurs physiques et climatiques. Contrairement aux idées véhiculées sur le pastoralisme Mbororo, il n'est pas une activité soumise à un nomadisme inévitable. Ce sont des conditions climatiques austères qui en constituent le principal facteur explicatif.

La quête d'une nouvelle citoyenneté

Les Mbororo installés aujourd'hui en RD Congo ne se considèrent ni Congolais ni étrangers. Leur mode de vie amène à penser qu'ils sont des citoyens du monde à la recherche des conditions climatiques favorables à la pratique du pastoralisme. Les frontières étatiques constituent pour eux une contrainte institutionnelle contre laquelle ils se mobilisent.

Certains d'entre eux estiment que leurs enfants ayant vu le jour en RD Congo sont d'office des Congolais. Cette conception de la citoyenneté n'est pas conforme à la loi sur la nationalité, qui elle, insiste sur le lien de sang[7]. On est Congolais parce que l'un des parents est Congolais.

Pour une solution régionalement intégrée

La République démocratique du Congo n'est pas un pays à tradition de transhumance pastorale. Cependant, sa position géophysique et géo-écologique attirera de plus en plus les éleveurs en quête de pâturages pour leur bétail. Voilà qui nécessite une harmonisation des politiques et une mise en commun des stratégies dans un cadre institutionnellement régional. En se mettant d'accord sur l'encadrement de la transhumance pastorale dans cette région, les États de la région pourront, par ce geste, témoigner de solidarité et d'esprit communautaire.

La dimension politique est appelée à acquérir une importance cruciale, car les changements climatiques qui affectent les régions d'où partent les éleveurs vont de plus en plus accélérer les migrations pastorales dans les prochaines décennies. Comme c'est souvent le cas, toute activité d'intégration ou de coopération régionale se heurte à de nombreux obstacles d'ordre technique, financier et humain qui, bien trop souvent, dépendent de la volonté politique des différents acteurs en présence.

Dès lors, cette intégration doit d'abord être fondée sur la définition claire d'une vision régionale du pastoralisme, sur la constitution d'une base de données et d'échanges de données à l'échelle régionale, sur la formulation concertée de stratégies communes pour un règlement durable des migrations pastorales.

Plus qu'un simple support aux activités pastorales, l'existence de cet espace régional est appelé à devenir un préalable, un facteur déterminant dans l'émergence d'un pastoralisme mieux encadré, faiseur de prospérité et de progrès dans une région en proie aux conflits armés récurrents.

La définition des politiques et des programmes régionaux dans le secteur du pastoralisme peut former une toile de fond sur laquelle il est possible de faire avancer l'intégration régionale.

Avec la rareté des pâturages, la transhumance est devenue un ingrédient permanent des conflits communautaires qui prennent des formes variées. Ceci exige qu'on comprenne les contextes ainsi que les contraintes de chaque pays et que l'on avance vers une harmonisation des législations nationales pour une bonne prise en compte des dynamiques nationales dans le cadre d'une intégration mieux comprise.

Le combat engagé pour protéger ou sauver l'environnement, renverser les processus de désertification est donc aussi une course contre le temps. Pour enrayer la dynamique de destruction des ressources naturelles, pour tenter de restaurer des équilibres écologiques viables dans les pays de la région, on ne dispose le plus souvent que de deux ou trois décennies, la plupart des indicateurs semblant bien montrer que la poursuite des tendances actuelles conduira, dans ces délais, à des irréversibilités ou à des dégradations catastrophiques.

Notes

1. L'année de l'arrivée des Mbororo en République démocratique du Congo est matière à controverse. Certains considèrent que les Mbororo sont sur le territoire congolais depuis les années 1980. D'autres encore avancent les années 1990.
2. En moins de cinquante ans, le lac Tchad a perdu plus de 90% de sa superficie totale. Il est passé d'une superficie de 260 000 km2 à 2 500 km2. La situation est inquiétante, surtout pour les pays qui sollicitent le lac Tchad pour la bonne marche de leurs stratégies économiques. Il subvient au besoin en eau des habitants du Tchad, du Niger, du Cameroun et du Nigéria. De nombreux scientifiques se sont attelés à l'analyse des causes de l'assèchement du lac Tchad. Selon les résultats de leurs études les plus récentes, le réchauffement climatique et l'avancée du désert, entraînant la rareté des pluies, sont les premières causes de cette situation. À ces phénomènes climatiques s'ajoute l'utilisation abusive de l'eau du lac pour de nombreux projets d'irrigation. En complément de ces analyses scientifiques, le ministre tchadien de l'Environnement et des Ressources Halieutiques, Hassan Terap, explique le phénomène par la prolifération de papyrus aux abords du lac. Des plantes à forte consommation hydrique.
3. Il s'agit bien des cas qui concernent le regroupement familial, les visas professionnels, diplomatiques, etc.
4. Cette classification consistait à distinguer les terres domaniales et les terres enregistrées d'une part et les terres indigènes de l'autre.
5. Voir la constitution, article 9, qui dispose que le sol et le sous-sol appartiennent à l'État. Les conditions de leur concession sont fixées par la loi, qui doit protéger les intérêts des populations locales.
6. Le nombre des milices armées est aujourd'hui assez imprécis en raison de la disparition et de la réapparition sous de nouvelles formes d'anciennes milices engagées dans le processus de démobilisation et de réinsertion. La radio Okapi qui avait mené une enquête dans la région avait recensé au moins 44 groupes armés.
7. La loi sur la nationalité insiste sur le fait que la nationalité de naissance s'acquiert du fait que l'un des parents est lui-même congolais.

Références

Bebbington, A., 1999, « Capitals and Capabilities : A Framework for Analyzing Peasant Viability, Rural Livelihoods and Poverty », *World Development*, 27 (12), p. 2 021-2 044.

Ela Ango, P., 1996, « La coopération militaire franco-africaine et la nouvelle donne des conflits en Afrique », *Relations Internationales et Stratégiques*, n° 23, automne 1996, p. 178-186.

FAO-ADAPT, 2011, Programme-cadre sur l'adaptation au changement climatique.

Haas, Hein de, 2010, « Migration and development : a theoretical perspective », *International migration review*, volume 44 (1), p. 1-38.

Haas, Hein de & Hassan El Ghanjou, 2000, *General Introduction to the Todgha Valley (Ouarzazate, Morocco) : Population, Migration and Agricultural Development*, IMAROM working paper series n° 5, Amsterdam, University of Amsterdam.

International crisis group, 2014, *Afrique centrale : les défis sécuritaire du pastoralisme*, Rapport Afrique n° 215, Belgique.

Kabbanji, L., 2011, « Vers une reconfiguration de l'agenda politique migratoire en Afrique de l'Ouest », *Études internationales*, 42 (1), p. 47-71.

Lazarev, Grigori, 2009, *La gouvernance territoriale et ses enjeux pour la gestion des ressources naturelles. Des approches novatrices pour lutter contre la désertification et la dégradation des terres et des eaux*, UNCCD, Secrétariat de la Convention des Nations Unies sur la Lutte contre la Désertification.

Lee, E., 1966, « A theory of migration », *Demography*, 3 (1), p. 47-57 (version française dans Piché, 2013, ch. 4).

Mabogunje, A., 1970, « Systems approach to a theory of rural-urban migration », *Geographical Analysis*, 2 (1), p. 1-18 (version française dans Piché, 2013, ch. 6).

Mpoyi Mbunga, A., B. Nyamwoga, F. Kengoum & F. Kabamba, 2013, *Le contexte de la REDD+ en RDC : causes, agents et institutions*, CIFOR.

Mpoyi Mbunga, A., *Le statut des droits des communautés locales sur les ressources naturelles en RDC*, rapport de consultation, 2011.

Nkoy Elela, D. (dir.), 2007, *Les migrations transfrontalières des Mbororo au Nord-est de la République démocratique du Congo. Etude de cas au Haut-Uélé et au Bas-Uélé*, Programme Grands-Lacs Pax Christi.

Parlement européen, 2008, *Les migrations climatiques*. Actes de la conférence du 11 Juin 2008 au parlement européen.

Piché, V., 2012, *Les théories migratoires contemporaines aux prismes des textes fondateurs*, Population-F, 68 (1), p. 153-178.

Portes, A., 1981, « Modes of structural incorporation and present theories of labor migration », *in* Kritz M. M., C. B. Keely & S. M. Tomasi (eds.), *Global Trends in Migration : Theory and Research on International Population Movements*, New York, The Center for Migration Studies, p. 279-297.

République démocratique du Congo, ministère de l'Environnement, Conservation de la Nature et Tourisme, 2012, Rapport national sur le développement durable en République démocratique du Congo.

République démocratique du Congo, 2011, Loi portant principes fondamentaux relatifs à l'agriculture, Journal officiel.

République démocratique du Congo, 1973, Loi foncière, articles 387, 388, 389, Journal officiel.

République du Tchad, ministère tchadien de l'Environnement et des Ressources halieutiques 2010, « Stratégie Nationale et plan d'actions pour la mise en œuvre de l'Initiative grande muraille verte », N'Ndjamena.

Tubiana L., F. Gemenne & A. Magnan, 2010, *Anticiper pour s'adapter, Le nouvel enjeu du changement climatique*, Pearson Education France.

3

Changements climatiques, genre, et inégalités sociales : les praticiennes de la médecine et de la pharmacopée traditionnelle en milieu urbain au Burkina Faso

Claudine V. Rouamba Ouédraogo & Natéwindé Sawadogo

Introduction

Les modèles d'urbanisation des villes africaines offrent des opportunités de compréhension de processus urbains plus larges. Ils permettent notamment, d'analyser des processus structurels tels que la relative reproduction sociale de la position vulnérable des femmes. L'approvisionnement des services de soins en médecine et pharmacopée traditionnelle à Ouagadougou, au Burkina Faso, constitue un des cas concrets révélateurs de ces processus. En effet, l'une des caractéristiques de l'urbanisation est sa tendance à reléguer les processus sociaux ruraux au second plan, voire à provoquer leur disparition totale. À Ouagadougou, ainsi que dans d'autres villes africaines, les praticiennes de la médecine et de la pharmacopée traditionnelle, comme nombre de catégories sociales, avaient jusque-là trouvé dans ces marges de la vie urbaine des niches protégées pour leur insertion urbaine à travers cette offre de services originellement ruraux. L'objet de cet article est d'explorer, à travers un exemple burkinabé, les changements climatiques, qui ont un impact direct et décisif sur les ressources naturelles (végétales, animales, minérales) et qui peuvent remettre en cause des acquis importants liés aux conditions de vie des femmes en milieu urbain africain.

Un tel propos présente un défi à la sociologie telle qu'elle a été pratiquée depuis la fin des années 1960. Depuis cette date en effet, l'engouement pour

les approches interprétatives a eu pour conséquence de prévenir l'attention
des sociologues sur la valeur heuristique des approches structuralistes (Blume
1969). Accusées d'avoir exproprié l'agenceité de l'individu en faveur de structures
sociales qui le domineraient, les approches qui se proposent de comprendre le
comportement ou la situation de l'individu en rapport avec d'autres processus
extérieurs humains ont graduellement perdu de leur attrait et ce jusqu'à une
période récente (Abbott 1988 ; Elias 1978 ; Hughes 1993 ; Parsons 1991). Quant
à l'idée de convoquer des phénomènes non sociaux pour éclairer des processus
sociaux, c'est plutôt en sociologie des sciences que le débat a eu lieu, qu'en
sociologie générale *per se* (Latour 1987). Que les sociétés humaines participent à
un ordre moral dont on ne peut trouver le parallèle à aucun niveau non-humain
n'est pas en débat. Le propos est plutôt, que l'individu est un être de relations
impliquant aussi bien d'autres êtres humains que des objets, et dont la dynamique
générale d'existence est instrumentale, sans que l'individu en question en ait pour
autant le contrôle complet. C'est du moins ce que propose l'écologie humaine.
En substance, l'écologie humaine est une approche sociologique développée par
le sociologue américain Robert Park (1936) pour étudier les dynamiques urbaines
et le changement social en général. L'approche est construite autour de trois
concepts : dominance, compétition, et succession. Le concept de dominance se
réfère aux caractères généraux de processus sociaux géographiquement situés. Celui
de compétition désigne le mécanisme par lequel la sélection s'opère pour donner
à ces processus leurs caractères relativement homogènes. Quant au concept de
succession, il traduit la continuité de ces processus sociaux ou leur remplacement
par d'autres de nature différente, résultant du processus de compétition entre
processus sociaux concurrents. Par exemple dans une ville capitaliste, la valeur
des quartiers est l'inverse de leur distance par rapport au centre qui abrite le
quartier commercial. En d'autres termes, plus un quartier s'éloigne du centre,
moins sa valeur est élevée. La position de chaque quartier est la résultante de
la compétition entre résidents pour le centre. La succession entre résidents de
même qualité, soit à travers des migrations, soit par simple reproduction (ou
adaptation) des mêmes processus, assure la continuité d'une structure urbaine
donnée. De même, plus la ville s'étend, plus la valeur du centre et des localités
proches du centre augmente de façon relative. Étant donné l'augmentation de la
valeur, il faut plus de ressources à un habitant pour y rester pendant longtemps ; à
défaut de pouvoir s'adapter, l'individu doit changer d'habitat pour être remplacé
par un autre qui en a les ressources. Que les villes contemporaines soient pluri-
centriques, ou que les localités périphériques acquièrent plus de valeur ne remet
pas fondamentalement en cause les principes de dominance, compétition et
succession de l'écologie humaine. En fait l'approche a même acquis un caractère
plus général, déconnecté de l'espace physique, pour s'appliquer au système social
général en termes d'occupation de positions sociales, notamment professionnelles
(Abbott 1988).

Les conséquences des changements climatiques sur les opportunités économiques des praticiennes de la médecine et de la pharmacopée traditionnelle à Ouagadougou peuvent être analysées de la même manière. En effet, il est important d'étudier les implications de la raréfaction des espèces naturelles médicinales sur la recomposition des métiers de la médecine traditionnelle en ville. Basé sur des enquêtes quantitatives et qualitatives, cet article met en parallèle changements climatiques et sources de revenu des femmes urbaines, afin d'analyser leurs implications sur les conditions de vie des femmes et des enfants. L'article est structuré en cinq sections. La première décrit la configuration sociale de la ville de Ouagadougou. Dans la seconde section, cette configuration est mise en rapport avec les besoins médicaux potentiels qui soutiennent les métiers de tradi-praticien. L'émergence des conditions favorables à un marché des soins infantiles et maternels traditionnels est analysée dans la troisième partie. La quatrième analyse la dynamique de la biodiversité au Burkina Faso. La dernière partie analyse les potentielles menaces qui pèsent sur ces opportunités en rapport avec les changements climatiques.

Ouagadougou : brève histoire sociale

Les besoins médicaux des patientes se rapportent à la structure sociale. Cette dernière est cause de leur existence, de leur continuité, de leur changement (Abbott 1988 ; Hughes 1993 ; Park 1936). Toute compréhension des processus écologiques doit se donner les moyens de comprendre le processus de composition-recomposition de cette structure sociale (Elias 1978). Par conséquent, il s'agit d'abord d'évoquer l'histoire sociale de la ville de Ouagadougou. Cette analyse se limitera à fournir les éléments les plus pertinents pour l'objectif de cette recherche. Toutefois, la recherche historique n'a pas encore pu nous fournir des certitudes sur les premiers habitants de la ville (*Wogdogo*) (Ki-Zerbo 1978 ; Simporé et Nacanabo 2006). Certaines sources avancent l'hypothèse qu'elle fut d'abord habitée par des Dogon (Simporé & Nacanabo 2006). En attendant d'autres éléments, la littérature actuelle établit que les *Ninsi* et les *Yonyonse* sont les autochtones de Ouagadougou (Dim Delobsom 1932 ; Halpougdou 1992). Par la suite, en 1495, ces deux groupes (pré-Dagomba) sont passés sous l'autorité des migrants Dagomba, les *Nakombse*, venus de Gambaga dans l'actuel Ghana, qui se sont d'abord installés à Tengkodogo à l'Est du pays (Skinner 1989). Ensemble, avec d'autres conquêtes de peuples voisins, ils ont formé à travers un long processus, les Royaumes mossi, avec pour capitale Ouagadougou, le *Moog-Naaba* étant leur grand souverain. La structure politique qui s'est développée à partir de ce processus et ses modifications ultérieures à la suite de l'arrivée de nouveaux groupes sociaux fournissent des éléments clés pour la compréhension de la structure sociale du royaume. Ce nouvel équilibre des pouvoirs reconfigure une structure sociale qui se manifeste par le développement d'un système de statuts distinguant les *Nakomse* (les détenteurs du pouvoir politique), la position dominante (avec ses propres stratifications internes), les *talse* (groupes sans lien de parenté avec le *naam*, pouvoir politique des Nakomse), les *tengembiisi* (fils de la terre : *Yonyonse*), et les *Yembse* (esclaves, capturés ou achetés). La plupart des membres du groupe des *Ninsi*, considérés comme agressifs, ont alors émigré de cette région centrale vers le Nord-Ouest, modifiant ainsi la structure sociale ancienne formée par les groupes pré-Dagomba. Le nouveau système politique qui s'est développé (et persiste jusqu'à nos jours) reflète le rang politique des différents groupes, bien que le système semble être plus fonctionnel que hiérarchique. La Cour du Roi des Moose (*Moog-Naaba*), qui est reproduite dans chacun des royaumes subordonnés, a donc à son sommet, le roi, autour duquel se manifeste la complémentarité entre le *tengembiisi* (fils de la terre : *Yonyonse*) et les détenteurs du pouvoir politique (*Nakomse*), y compris le roi et ses auxiliaires (*Talse, yembse*).

Par la suite d'autres groupes sociaux immigrèrent à Ouagadougou. D'abord, il y eut les *Yarse,* du groupe Mandé. Ils sont souvent assimilés aux Bambara et aux Dioula. Les Bambara sont arrivés à Ouagadougou au XVIᵉ siècle (Simporé 2009), grâce au commerce. Ils sont marchands et pratiquaient le commerce à longue distance (Ki-Zerbo 1978 ; Simporé et Nacanabo 2006 ; Simporé 2009). Ils ont ainsi contribué à relier, à travers ces échanges commerciaux, les différentes parties des autres royaumes mossi à la ville royale de Ouagadougou, puis Ouagadougou à la côte Atlantique et à l'actuel Mali (Sedogo 2006). De Ouagadougou, ils amenaient des tissus de coton vers le Mali ; puis ils ramenaient du Mali du sel et du poisson qu'ils vendaient tout au long de leur itinéraire jusqu'à Ouagadougou. Ils retournaient ensuite au Ghana où ils échangeaient les mêmes produits contre des noix de cola. Les noix de cola et le sel étaient fort appréciées par les Moose, car ils étaient utilisés dans diverses circonstances (mariage, funérailles, dons à la belle-famille, à la noblesse, etc.). Ce commerce à longue distance leur avait donné aussi le monopole de l'élevage et du commerce des ânes, qui s'était développé à la suite du commerce caravanier, dont ils avaient également le contrôle (Sedogo 2006). Monteil (1995:252), un explorateur français qui a visité le pays à la fin du XIXᵉ siècle, observait que les *Yarse* étaient caractérisés par « une réelle compréhension des affaires commerciales et d'une grande audace ». Leurs compétences littéraires, de même que leur accès à l'information, leur avaient permis de profiter de faveurs exceptionnelles de la part du souverain. Les souverains de Ouagadougou « ont également bénéficié plus ou moins de leurs pouvoirs magiques et de leurs conseils » (Sedogo 2006:101). Les *Yarse* sont des musulmans. S'ils excellaient dans le commerce qui est resté longtemps sous leur contrôle, les *Yarse* étaient aussi bons artisans dans le tissage. Depuis le XVIIᵉ siècle, les communautés *Yarse* se sont développées autour des marchés de villages importants du royaume (Kouand 1996 ; Skinner 1989 ; Audouin & Deniel 1978).

Ensuite les Haoussa sont arrivés. Venus en grande partie du Sokoto, les *Haoussa*, marchands musulmans, se sont installés à Ouagadougou au XVIIIᵉ siècle. Ils voyageaient par le même itinéraire que les *Yarse*. Cependant, contrairement aux

Yarse, les *Haoussa* étaient itinérants et s'installaient rarement en un lieu fixe pour une longue période : « Leurs caravanes passaient par le pays de Kano et les bords du Tchad, jusqu'à la côte de Lagos, pour traverser ensuite le Mossi, Yatenga, Douentza, Macina, et finalement atteindre Tombouctou. » (Monteil 1991:252) En plus de ces groupes, il y a les Peuls. D'origine berbère, les pasteurs peuls sont éleveurs et musulmans. Pendant la période précoloniale seuls quelques privilégiés possédaient du bétail (seuls les rois pouvaient en avoir). Nomades, les Peuls menaient une vie itinérante. Les chefs de Ouagadougou préféraient développer des relations contractuelles avec eux pour l'élevage de leurs bovins. Contrairement aux *Yarse*, les Peuls n'avaient « pas une part importante dans les affaires publiques » (Monteil 1991:253). D'autres groupes sont arrivés après ces derniers, tels que les griot artisans (Laobé), et d'autres artisans comme les *Marense*, les tanneurs (Zap-Ramba). Les cordonniers-tanneurs fabriquaient du matériel de guerre.

Les *Ninsi*, premiers habitants de Ouagadougou, pratiquaient le commerce, mais ils étaient encore dans une économie de troc. Il n'y avait pas de marchands spécialisés, de sorte que les échanges se faisaient en produits locaux. Les *Yonyonse*, étaient agriculteurs et éleveurs. Quant aux *Nakombse*, ils ont pratiqué diverses activités (agriculture, élevage, artisanat, etc., l'agriculture et l'élevage étant leurs activités dominantes). Cependant en dehors de l'administration de la cour du roi, l'activité principale des *Nakomse* était la guerre. La majeure partie de leurs revenus provenait des raids (Sedogo 2006). L'établissement des *Yarse* et des *Haoussa* au XVIᵉ siècle a eu un impact significatif sur la structure de l'économie. Le premier changement important est l'augmentation de la population de marchands spécialisés. Les *Yarse* étaient installés dans les centres importants du pays. Les *Haoussa*, bien moins enclins à de longues périodes d'implantation, sont venus en grand nombre à Ouagadougou. Parmi la population autochtone, beaucoup sont passés de l'agriculture et de l'élevage au commerce (Monteil 1995). Cela s'est traduit par le développement d'une économie de marché impliquant l'usage des cauris comme monnaie. Le développement de l'économie de marché, avec son corollaire, la diminution de la population impliquée dans l'agriculture et dans l'élevage en raison de la réduction de l'espace disponible pour de telles activités, fit de Ouagadougou un important centre urbain. L'implication de la population autochtone dans les activités commerciales a favorisé leur conversion à l'islam (Monteil 1991 ; Binger). Cela n'a pas concerné seulement les gens ordinaires, mais aussi la noblesse (Englebert 1996 ; Skinner 1989). Les groupes sociaux aux conditions sociales modestes ont trouvé dans le commerce et dans l'islam une alternative pour leur émancipation. Par ailleurs, le développement économique a eu lieu pendant une période d'insécurité dans les pays voisins. Ouagadougou était donc devenu un refuge pour nombre de personnes de ces sociétés. En outre, le contexte d'insécurité avait rendu nécessaire la présence d'un poste de sécurité parmi les caraviniers (Monteil 1995). Un groupe d'individus en venait donc à tirer leurs subsistances des emplois indirects créés par le développement commercial et l'urbanisation. Une fraction relativement importante de la population autochtone était devenue moins dépendante de la terre pour leur subsistance et de la religion traditionnelle comme référent symbolique. Cela avait à son tour influencé la circulation des femmes, parce que si les autochtones animistes pouvaient donner leurs filles en mariage aux musulmans, l'inverse était exclu par ces derniers, sauf si le mari acceptait de se convertir à l'islam.

À la fin du XIXe siècle, tous ces groupes sociaux vont passer sous la domination d'un autre : l'État français. En effet, les Français conquirent le royaume du Yatenga en 1895, puis celui de Ouagadougou en 1896. Dès 1898 presque tous les groupes voltaïques sont passés sous l'autorité française (Suret-Canal 1977 ; Crowder 1968 ; Ki-Zerbo 1978). Ceux-ci ont été donc administrativement intégrés au grand ensemble créé par le décret du 16 juin 1895 définissant l'Afrique-Occidentale française. Cette dernière comprenait le Sénégal, le Soudan (actuel Mali), la Guinée, la Côte-d'Ivoire, le Dahomey (actuel Bénin), la Mauritanie, la Haute-Volta (actuel Burkina Faso) et le Niger (Brasseur 1997). Le processus d'intégration territoriale a pris seulement une dizaine d'années. Au lendemain de la conquête, la structure de la population se présentait de la façon suivante.

Figure 3.1 : Répartition des religions en fonction des groupes ethniques à Ouagadougou en 1909

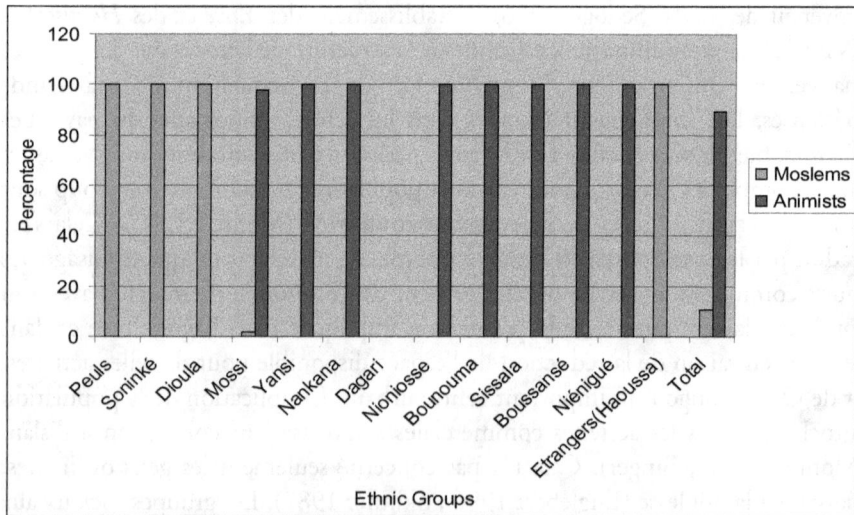

Source : Sawadogo, N. 2013

Ainsi le Burkina Faso (ancienne Haute-Volta) est devenu successivement partie du Soudan français (1894-1898), du Haut-Sénégal-Niger (1904-1919), de l'Afrique-Occidentale française (1895-1898), et une colonie indépendante (1919-1937 ; 1947-1958) (Massa et Madiega 1995 ; Becker, Mbaye & Thioub 1997). Le pays accéda à l'indépendance en 1960 et Ouagadougou devint la capitale du pays indépendant. La première constitution du pays fut adoptée en 1959. Cette constitution a été modifiée par la suite et adoptée par la Première République, proclamée le 5 août 1960. Mais six ans après, le pays passa sous un régime militaire (1966-1970), qui l'a conduit à la Seconde République (1971-1974). Puis le retour du pays à un régime militaire (1975-1977) fut suivi de nouveau par un régime constitutionnel à travers la Troisième République (1978-1980). Jusqu'en 1991, le pays était sous régime militaire (1980-1982 ; 1982-1983 ; 1983-1987 ; 1987-1991) connaissant différents destins. Le dernier recensement général de la population de 2006 a enregistré 14 017 262 habitants, dont 3 181 967 concentrés dans les villes et 10 835 295 en zone rurale. Le taux d'urbanisation était de 22,7 pour cent et le taux d'accroissement annuel moyen de la population était de 3,1 pour cent. Le pays a une superficie de 274 200 kilomètres carrés (RGPH 2006). Ouagadougou est la plus grande ville du pays avec 21 930 hectares, une population de 1 475 839 habitants, et 7,6 pour cent de la population taux de croissance annuelle (RGPH 2006). Selon le recensement, le taux d'analphabétisme de la ville reste élevé (40,1 %). Les principales religions sont la musulmane (57 %), la catholique (34,9 %) et la protestante (6,2 %) ; les animistes sont estimés à 0,4 pour cent seulement. Plus de 80 pour cent de la population de plus de 15 ans travaillent dans le secteur tertiaire, dont 52,6 pour cent dans l'artisanat, ou comme vendeurs ou personnel de service ; moins de 8 pour cent sont dans le secteur primaire, et 11,4 pour cent dans le secteur secondaire ; de cette population active, seuls 35,3 pour cent sont salariés ; les autres travaillent dans l'entreprise indépendante (46,3 %) ou les employeurs (2,8 %) (RGPH 2006). Cet aperçu de la recomposition de la ville de Ouagadougou dans la longue durée, permet de comprendre la genèse du métier de tradi-praticien que les femmes en sont venues à exercer et les défis auxquels elles sont confrontées dans le contexte actuel des changements climatiques.

Structure sociale et exclusion de la femme du marché des soins

C'est par rapport à cette structure sociale que l'on peut comprendre non seulement les besoins médicaux des patientes, mais aussi l'accès et la valorisation des compétences médicales des femmes et leurs implications structurelles. Comme les compétences, les besoins ne sont pas neutres et universellement interchangeables ; ce sont des construits sociaux qui résultent de la structure sociale (Hughes 1984). Dans la société traditionnelle moaga c'est la parenté qui fournit le cadre d'interprétation de la maladie de l'enfant, comme elle le fait pour les maladies de la reproduction de la mère. Le propos suivant d'un vieux dans un village à la périphérie de Ouagadougou l'illustre :

> Si un enfant a un an et demi, deux ans, il n'a rien ! Ce qui va lui faire mal ce sont ses dents. C'est ça seulement qui fatigue l'enfant ; sinon quand ça (un autre malaise) commence seulement... on part avec lui chez le devin [cité dans Sawadogo 2006:59]

Mais pour comprendre la parenté, c'est au mythe qu'il faut se référer. Le mythe est une production symbolique et active. Il constitue de ce fait une des conditions du fonctionnement de la structure sociale. Ainsi, comme pour les *Tallensi* étudiés par Fortes, pour les Moose, entre le monde sensible et le monde mythique, il n'y a pas de frontière. Il y a une circulation continue de l'un à l'autre, car « l'intérieur n'est pas juxtaposé à l'extérieur, comme des domaines séparés. Ils se reflètent l'un dans l'autre et ce n'est que dans ce jeu réciproque de miroirs qu'ils dévoilent leur contenu intime », (Fortes 1974). Cette relation se traduit principalement dans le culte des ancêtres. En effet, les Moose reconnaissent l'existence d'une divinité suprême, *Wende*, maître du cosmos. Skinner le qualifie de « divinité négligente » parce qu'on ne rencontre pas de culte qui lui soit destiné. En revanche, le culte des ancêtres occupa une place considérable dans leurs croyances religieuses. Les Moose croyaient en une implication régulière des esprits ancestraux dans les affaires des vivants. Mais leurs interventions sont ambivalentes. Ils sont tantôt bienveillants, tantôt malveillants. L'exercice de leur justice avait pour but de renforcer les normes en vue de la pérennité de l'ordre social. Le statut d'ancêtre n'est pas indifféremment attribué à tous les morts. Il s'acquiert sur la base de critères gérontocratiques et éthiques. Il y a une relation intime entre le culte des ancêtres et la représentation de la personne et de la mort. Ainsi que le note Cassirer, « là où règne cette croyance, l'individu ne se sent pas seulement liés aux ancêtres de sa tribu par le processus continu des générations : il se sent identique à eux. Les âmes des ascendants ne sont pas mortes : elles existent pour s'incarner de nouveau dans les descendants, et pour se renouveler sans cesse dans les générations futures » (Cassirer 1972:208). Cette « participation mythique de la personne » se révèle principalement dans la représentation collective de quatre de ses composantes : le *sègré* (esprit titulaire ; de façon littérale : rencontrer), le *siiga* (principe vital), le *kiima* (ame), le *kinkirga* (génie), et le nom individuel. Le *sègré* symbolise l'hérédité. C'est un ancêtre : « l'ancêtre familial détermine qui est censé revenir, c'est-à-dire prendre en charge l'enfant nouvellement né, pour l'aider, le protéger, et le surveiller dans sa vie terrestre », (Ouédraogo cité dans Poulet 1970:113). Le *siiga* quant à lui constitue le principe vital corporel : « C'est l'âme végétative qui anime tout être, les hommes comme les animaux et les végétaux », (Badini 1970:802). Le *kiima* est une composante dynamique et il dispose d'une existence propre. Les Moose concevaient les ancêtres sous cette dénomination et leur adressaient des cultes. Son statut dépend de l'importance sociale de la lignée de l'individu. Cette composante rejoindrait les autres âmes du lignage à Pilimpikou (un village mythique mossi). Le *kinkirga* symbolise l'individualité. Les *kinkirsi* (génies) pluriel de *kinkirga* peupleraient tout l'univers, mais la tradition fait des discontinuités de la nature (marigot, montagne, terrains dénués) leur lieu privilégié. Leur importance apparaissait essentiellement dans le cadre de la conception. C'est ce cadre mythique qui fournit les catégories d'interprétation des évènements (malheureux ou heureux) qui se produisent pour l'enfant, et de l'expérience génésique de la femme.

Q. – Avant, quand un enfant tombait malade, à quoi ses parents pensaient immédiatement ?

R. – C'est comme je l'ai déjà dit. Je vous avais dit que quand un enfant était né, quand il commencera à marcher, il pleure. Et quand il n'a pas encore de nom, on dit que c'est ceci ou cela. Si toi-même tu fais une nuit blanche et tu sors pour chercher les causes, si on te dit que l'enfant pleure parce que c'est ceci ou cela, ne vas-tu pas suivre ? Hein ! si vous revenez faire ce qu'on vous a montré et l'enfant cesse de pleurer, vous allez savoir que l'affaire est vraie ! Ce n'est pas faux (rire). C'est pourquoi on ne peut pas expliquer tous les secrets. On ne peut pas... Hein !! Il se peut que le problème vienne de toi-même ou de ton père, hein ! Mais vous-mêmes vous ne savez pas. Si vous partez demander à quelqu'un il vous montre.

Q. – Demander à qui ?

R. – Il y a des gens qui sont des savants et si vous partez leur demander, ils diront que c'est à cause de ceci ou de cela que l'enfant est ainsi. S'il souffre d'une autre maladie ils diront que ceci ou cela : vous voyez ? C'est pourquoi les choses d'avant ne sont pas les mêmes que celles d'aujourd'hui. [cité dans Sawadogo 2006:59-60]

Ce cadre d'interprétation de la maladie de l'enfant suggère des types d'expertise qui se confondent avec la structure patriarcale de la société. En interprétant les maladies de l'enfant comme des résultantes du dysfonctionnement de la parenté, le discours valorise des types de relations impliquant exclusivement des hommes. En tant qu'acte religieux ce sont les prêtres de la communauté qui sont investis de l'intercession légitime auprès des ancêtres en faveur de l'enfant ou de la mère. Dans la société moaga, trois catégories de prêtres sont investies de cette légitimité selon le niveau d'intervention. Il y a le *Yirsoba* et le *Buudkasma*. Ce sont des quasi-prêtres, en ce sens que leur fonction se limite au niveau minimal du lignage qu'est la famille. Le chef de famille (ou selon le cas, l'aîné de la grande famille) intercède auprès de leur père défunt pour expier une faute ou demander une aide en faveur de l'enfant dont il est le protégé. Ensuite au niveau maximal du lignage il y a le grand prêtre, qu'est le *tengsoba* (ou le chef selon le cas). Il intercède auprès des ancêtres pour des questions de santé d'intérêt général, en intercédant auprès des aînés du lignage. Il est aussi le dernier recours lorsque les intercessions auprès des aînés au niveau minimal de la famille s'avèrent infructueuses. Dans tous les cas, les femmes sont structurellement exclues de l'accès à ce rôle et à l'expertise correspondante. Les rites font suite aux recommandations d'un devin. En général c'est un homme. On pense qu'il a une double vue qui lui permet de communiquer avec les êtres du monde invisible afin de prescrire les sacrifices nécessaires pour expier les mauvais esprits ou résorber la faute morale envers les ancêtres. Néanmoins le *kinkirbaga* est généralement une vielle femme. Les *kinkirse* l'adoptent comme leur mère. En dehors de ces prêtres il y a d'autres acteurs qui interviennent dans la gestion de la maladie, parmi lesquels il peut y avoir des femmes. Ce sont, notamment, les herboristes. Cette catégorie de praticiens a généralement une interaction physique avec la personne malade. Si la maladie est déjà identifiée, le praticien ou la praticienne donne des médicaments pour le traitement ; sinon, le malade est sommé de consulter un devin. Ce type de praticien ne pratique pas la divination (Bonnet 1988). Cette connaissance est acquise par la révélation par un ancêtre, par esprits de la terre, ou bien transmise par les parents voire achetée ; elle peut également avoir été construite à travers une expérience de la maladie (Bado 1995 ; Bado 2006). Il y a également les accoucheuses traditionnelles qui sont des vieilles femmes atteintes par la ménopause. Une accoucheuse n'est pas choisie par le village :

> Elle devient accoucheuse par son propre choix, après avoir assisté pendant plusieurs années une autre accoucheuse qu'elle remplace lorsque celle-ci devient très vieille ou meurt ; « l'obligation » sociale principale est d'avoir donné naissance elle-même et d'avoir atteint la ménopause. (Bonnet 1988:41)

Les avortements spontanés sont traités par les vieilles femmes de la famille de la femme avortée. Après l'accouchement, les femmes âgées de sa famille prennent en charge la mère et son enfant. Toutefois, l'accoucheuse lui montre comment prendre soin de l'enfant immédiatement après la naissance, soins que la femme répète jusqu'à ce que les impuretés soient nettoyées. Avant d'être mère, la femme apprend son rôle de mère. Elle le fait dans le cadre de son éducation de fille ; si elle se marie et donne naissance, l'accoucheuse et les femmes âgées de la famille lui montrent comment prendre soin d'un bébé. Ses connaissances se développent plus ou moins en fonction de l'état de santé de sa progéniture (Badini 1970 ; Erny 1968 ; Erny 1999 ; Bonnet 1994 ; Fortes 1959 ; Lallemand 1971 ; Rabin 1979). Elle peut devenir accoucheuse, mais aussi herboriste selon sa propre histoire en tant que mère, mais aussi selon ses attaches familiales. Si les kinkirsi l'adoptent comme mère elle peut devenir devineresse, mais jamais elle ne peut être prêtre. Si elle est herboriste, son expertise est généralement limitée aux soins maternels et infantiles.

L'immigration des groupes musulmans va introduire une compétition institutionnelle à Ouagadougou. En effet, en principe l'islam est opposé à la religion traditionnelle mossi. L'islam est une religion monothéiste, qui défend l'unité et la transcendance d'un Dieu unique. L'islam conçoit que sur la terre il y a deux anges gardiens autour de chaque personne, mais aussi de mauvais anges. Contrairement aux anges gardiens, qui sont créés à partir de la lumière, les mauvais anges sont créés à partir de feu. Ils sont appelés génies. Les génies sont invisibles, astucieux et malveillants. Ils peuvent attaquer les êtres humains et les tourmenter. Contrairement à certains Kinkirsi dans la croyance religieuse mossi, la possession par les génies chez les musulmans (pluriel de djinns dans la langue maure) est toujours négative. Ce sont ces catégories religieuses et magiques qui offrent aux musulmans le cadre de leur interprétation des évènements qui surviennent dans la vie de l'enfant ou de sa mère. Il existe des différentiations distinguant les fonctions religieuses et temporelles entre les experts en médecine islamique. Dans tous les cas, dans la société moaga, l'islam consacre la défonctionnalisation totale de la femme mossi en matière de soins, n'autorisant que les compétences acquises en tant que mère.

Avec la colonisation les savoirs médicaux de tous ces groupes sont devenus illégitimes. « L'influence de la médecine occidentale, écrit Sakanlé, a suivi en Afrique, les chemins de la colonisation» (1969:29). Au sein de la colonne d'attaque qui a repris le pays Mossi, il y avait un médecin. En effet, en dehors des médecins en mission d'exploration, les premiers services médicaux modernes étaient fournis par les médecins des postes militaires des colonnes d'attaque. Il y avait aussi des centres de santé établis par les missionnaires. La première équipe missionnaire est arrivée au Burkina Faso le 22 janvier 1900, immédiatement après la conquête ; le premier centre de santé de Ouagadougou a été fondé en 1913, peu de temps après l'arrivée des premières sœurs en décembre 1912 (Monne 1999). Ce que pense Perrot en tant que botaniste, lors d'une mission au Burkina Faso, reflétait une attitude générale parmi le personnel colonial de santé, qui a cherché à établir un monopole sur les questions de santé dans la colonie. Il a ainsi indiqué que :

> À Ouagadougou, comme ailleurs, on a pris soin de l'hygiène individuelle générale de l'indigène, et l'hôpital qui a été géré à l'époque par le Dr Lairac et un autre médecin adjoint était des plus intéressants et des plus actifs. Une infirmière avec des sages-femmes auxiliaires indigènes ou métisses fournissaient également des services indiqués : les mesures de prévention commençaient à disparaître chez les Noirs ; il est cependant quelque chose à souligner et des plus agréables pour l'avenir ; il est permis de prévoir si ce n'est la suppression des matrones et leurs pratiques cruelles et dangereuses, du moins leur réduction notable, alors que pénètrent, dans la masse, les principes fondamentaux de l'hygiène préventive et de la propreté dans la prise en charge des femmes donnant naissance. Tout ce qui peut permettre la conservation de la race et la réduction de la mortalité infantile doit être poursuivi avec méthode et si nécessaire avec la dernière énergie. (Perrot 1927-28:82)

Les règles d'octroi du permis d'exercer excluent automatiquement les praticiens autochtones. Le premier article du décret de 1936 sur la pratique de la profession d'herboriste, déclare que :

> Nul ne peut exercer la profession d'herboriste, ouvrir une boutique en AOF (Afrique-Occidentale française) s'il n'a pas plus de 21 ans et ne possède un diplôme d'herboriste décerné par le gouvernement français, à la suite d'examens soutenus dans les facultés ou écoles de l'État.

En Afrique-Occidentale française, jusqu'en 1898, la préoccupation était davantage la conquête militaire, de sorte qu'avant 1905 les 42 centres médicaux de la colonie étaient gérés par des médecins militaires, diplômés d'écoles médicales de la marine telles que Toulon (1725), Rochefort (1732) et Brest (1757). En 1898, le gouverneur général de la colonie et son inspecteur des services de santé ont fait campagne pour une innovation dans la pratique médicale, qui permettrait d'utiliser les nouvelles avancées en matière de découvertes pastoriennes au bénéfice du projet colonial. Ils ont recommandé l'hygiène comme moyen de promotion de la médecine préventive et sociale. Ce fut un événement majeur parce que c'était la première initiative d'une médecine de masse au niveau de l'Afrique-Occidentale française d'alors, un territoire créé par décret le 16 Juin 1895, englobant plus de dix millions d'habitants. Cette initiative a été suivie par l'institution de l'Assistance médicale indigène (AMI), le 8 Février 1905, un organisme de santé similaire à celui de l'Aide médicale gratuite (AMA) en France, qui avait été auparavant mis en place à Madagascar en 1896 (Bado 2006). Le règlement qui a institué l'AMI a ensuite été modifié, une première fois en Janvier 1907, et la seconde modification a pris effet à partir de 1912, définissant une politique de santé des soins et des conseils gratuits pour la population autochtone, les Européens et leurs familles dans les colonies. Mais en fait, ces derniers étaient davantage pris en charge que la population autochtone, dont les bénéficiaires se limitaient aux soldats et à leurs familles. Néanmoins, c'était une première étape. La mise en œuvre de ces mesures s'est avérée difficile parce que la médecine moderne connaissait peu les maladies tropicales à l'époque. En 1903, 142 médecins militaires au total sont disponibles pour toutes les colonies françaises, et ils ont été attribués aux troupes militaires plutôt qu'aux organismes de santé civils. La pénurie de personnel pour le programme et les centres de production de vaccins pour la vaccination mobile, a conduit à de nouvelles mesures pour le recrutement de médecins civils afin de remplacer les médecins militaires pour qu'ils puissent rejoindre les troupes. La campagne de recrutement dans les facultés de médecine françaises a donné lieu à cinq demandes seulement, un nombre qui ne répondait pas aux besoins du gouvernement de l'AOF. Les médecins militaires ont donc été maintenus pour l'Assistance médicale indigène, et le décret de 1907 modifia celui de 1905 afin de permettre le recrutement d'aides-médecins pour compléter le groupe professionnel des infirmiers autochtones, essentiellement composé de militaires, créé en 1889. Un groupe professionnel d'aides-médecins a donc été créé en 1906. Les médecins coloniaux se plaignirent de la mauvaise performance de cette catégorie de professionnel de la santé, comme ils l'avaient fait pour les infirmières autochtones. Les plaintes étaient en partie justifiées, mais cela était en réalité lié à des questions de rivalité professionnelle. La résistance des administrateurs de la santé coloniale aux plaintes des médecins, en raison de leurs préjugés raciaux, a pris fin avec le début de la première guerre mondiale. Les médecins militaires ont été réaffectés aux troupes, et, plus important, les dirigeants politiques dans les colonies demandèrent plus d'égalité dans les droits civils, se fondant sur la participation de leur peuples aux cotés de la France.

La première école de médecine a donc été créée à Dakar en 1918 pour les colonies de l'Afrique occidentale française. Dix-neuf médecins du Burkina Faso ont été diplômés à l'école entre 1927 et 1953. D'autres écoles de formation en santé ont plus tard été établies au Burkina Faso : les écoles d'infirmières (1931), e Service de la prophylaxie et de la trypanosomiase (1932), l'école-Jamot de la maladie du sommeil (1937), le Service général mobile d'hygiène et de prophylaxie 1939), l'école de l'Assistance médicale indigène (1948), l'école de l'Assistance médicale africaine (1958). Le recensement de 1971-1972 a trouvé 5 hôpitaux, 13 dispensaires de ville, 178 dispensaires ruraux, et quelques centres médicaux ; e ratio de médecin par habitant était de 1/50 000, puis 1/200 000 cinq ans plus ard. Chacune des catégories de soignants dans les formations anciennes a son représentant moderne dans le système de santé colonial.

Émergence d'une niche professionnelle informelle pour les femmes

« Pour survivre, une institution doit trouver une place dans le style de vie des gens, ainsi que dans leurs sentiments » (Hughes 1984:11). D'un point de vue écologique, on s'aperçoit que la compétition conduit à un déclin graduel de certains besoins ayant pour corollaire celui des acteurs dont l'activité dépendait de ces besoins, alors que les besoins qui requièrent les services des tradi-praticiennes de santé augmentent. En effet, les services des prêtres sont devenus au fil du temps moins pertinents. Ces services dépendent de besoins dont la base est religieuse, elle-même fondée sur les règles de la parenté. Le contrôle de la circulation des femmes par le lignage est ainsi une condition nécessaire à la maintenance du système. Avec l'urbanisation et l'apparition des nouvelles religions le système traditionnel a perdu le contrôle religieux du mariage. Or sans ce contrôle religieux, le système médical traditionnel devient inopérant.

Un autre processus consiste en ce que, bien qu'administrativement et légalement, le système de santé moderne se soit octroyé le monopole de la santé, il présente des déficiences. A Ouagadougou, l'offre publique de santé est, au moins géographiquement, conforme aux normes internationales. Tout comme l'assistance médicale publique, la pratique privée existait dans les colonies et s'est poursuivie jusqu'à la fin des années 1960, sans nouvelles réglementations officielles. Selon Ouédraogo (2010), il s'agissait d'une pratique officieuse des médecins militaires français expatriés et établis entre 1966 et 1968. Les consultations étaient effectuées à l'hôpital dans leur bureau, en dehors du temps de travail officiel, et les frais laissés à l'initiative des médecins-expatriés. La pratique a ensuite été autorisée dans un règlement par le ministre des Finances de l'époque. Selon Douamba (2010:6) :

> Il faut reconnaître que, à l'indépendance, le projet colonial a été poursuivi parce que non seulement les agents nationaux de haut niveau étaient rares, mais ils étaient aussi jeunes diplômés et ne pouvait pas avoir accès à des postes de responsabilité. Tous les administrateurs, tous les chefs de services hospitaliers étaient, à quelques exceptions près, les agents de la coopération française, des militaires et du personnel civil. La politique de santé a donc été entièrement conçue et développée par les accords de coopération, et les officiers supérieurs revenant de formation ont été ulcérés d'être marginalisés.

La médecine traditionnelle était encore illégale. Mais les troubles civils qui on éclaté en 1966 au Burkina Faso, ont entraîné des modifications importante par le nouveau régime militaire. À partir de 1966, les autorités politiques « on commencé à grignoter les prérogatives des acteurs de la coopération étrangère e de se connectés à des organisations internationales telles que l'OMS » (Douamba 2010:6). Il semble que la baisse du taux de médecins par habitant de 1/50 000 er 1971-1972 à 1/200000 cinq ans plus tard, reflète ces changements, ainsi que le vide créé par la retraite de la génération de la médecine coloniale. Cependant, elle a également soulevé de nouveaux défis. Les élites ont hérité de ce système de santé et maintenu une gratuité des soins pour tous jusqu'à la fin des années 1980. Dan l'intervalle, la stratégie de santé primaire de l'OMS de 1973, remaniée plus tard grâce à l'Initiative de Bamako en 1987, a permis d'améliorer l'accès des populations rurales aux soins. La stratégie des soins de santé primaire a entraîné non seulemen la réhabilitation relative des guérisseurs traditionnels et en particulier des matrones mais aussi l'émergence des différentes catégories de personnel de santé au sein du système de santé. L'école privée d'infirmières, qui a hérité de l'école coloniale paramédicale de formation (AMI ; AMA), est devenue la seule formation du personnel de santé en 1963 ; elle a été transformée en une école nationale des infirmières et des sages-femmes, et a formé la plupart des professionnels de la santé pour la mise en œuvre de la politique de soins de santé primaire. L'école de médecine nationale a été ouverte en 1981 ; il y a actuellement quatre écoles de médecine dans le pays, avec seulement deux hôpitaux universitaires. Toutefois, les mesures de déréglementation, qui ont suivi les crises économiques de 1972-1974, quoique bloqués par le régime révolutionnaire (1983-1987) au Burkina Faso, se sont généralisées dans les années 1990. Le Burkina Faso a fait son entrée dans le marché en 1991, consacrant ainsi l'ouverture d'un nouvel espace dans le système national de soins de santé. La pratique médicale a été libéralisée et les différents domaines de pratique professionnelle ont été plus tard définis dans le Code de la santé publique par le n° 23/94/ADP de la loi du 19 mai 1994. La loi n° 034/98/ AN de l'hôpital/en mai 1998 organise le système de santé en définissant les différentes catégories de services de santé. Le décret du 19 juillet n° 398/PRES/ PM/MS 2005 complété respectivement par le décret interministériel et ministériel des règlements n° 2006/MS/MCPEA/MFB du 16 juin 2006 n° 200-060/MS/ CAB, a défini les conditions d'octroi de licences professionnelles, et l'ouverture et le fonctionnement des centres de soins de santé privés. Le nombre de centres de santé privés est passé de 58 en 1990 à 250 en 2000, puis 380 en 2009 (Ouédraogo 2010). Toutefois, le coût élevé des médicaments, conjugué à une administration peu accueillante et des services non adaptés ne favorisent pas la demande.

Un autre facteur favorable à la création de niche professionnelle est la reconnaissance officielle de la médecine traditionnelle. À propos de cette dernière en effet, « Après une léthargie vers 1960-1970, la fin des années 1970 a vu la mise en vigueur de l'ordre n° 70-68 bis/PRES/MSP/AS de décembre 1970 relatif au Code de la santé publique et les règles de son application ; cette loi tolère la médecine traditionnelle. » (Bognounou & Guinko 2005:33)

En 1994 la n° 23/94/ADP la loi sur le code de la santé publique :

> Reconnaît la médecine traditionnelle et la pharmacopée traditionnelle comme l'une des composantes du système national de santé […] En dehors de leurs activités traditionnelles d'évangélisation, l'une des caractéristiques communes aux églises chrétiennes de différentes confessions établies au Burkina Faso, est qu'elles apparaissent de plus en plus comme des lieux de soins de la maladie. (ibid.:34)

D'autre part il y a le fait que le contexte culturel et économique est favorable aux activités des praticiennes de la médecine traditionnelle. L'exode rural contribue à donner à la ville une composition relativement rurale. De plus, une part importante des ménages urbains est pauvre. De même, il existe maintenant une génération de femmes nées ou grandies en ville et en ayant reçu l'éducation, qui s'éloignent de l'éducation féminine telle qu'elle se déroule en milieu rural, mais que leur situation économique écarte des soins modernes. Toutes ces catégories de citadins constituent un marché. Les statistiques sont partielles. À ce jour, même au niveau national, il n'existe pas encore de recensement exhaustif. En 2004, le ministère de la Santé a estimé qu'il y avait plus de 30 000 « guérisseurs traditionnels » dans le pays, dont environ 3 000 à Ouagadougou, 1 500 à Bobo Dioulasso, la deuxième ville du pays, et 600 dans chacune des autres provinces. Cela donnait un ratio de 1 « guérisseur traditionnel » pour 500 habitants. Cela signifie que les 263 tradipraticiens interrogés dans le cadre de cette recherche représentent seulement une petite partie. Le recensement de la Direction régionale de la santé du Kadiogo (Ouagadougou) a enregistré 640 « guérisseurs traditionnels » en janvier 2010. Ce chiffre était de 278 en 2006. Nos propres enquêtes quoique partielles ont permis d'enregistrer 263 tradipraticiens, soit 111 hommes et 152 femmes. Seuls 94 des 263 enquêtés déclarent savoir lire et écrire (10 alphabétisés dans une langue nationale, 40 en arabe, 44 en français). Néanmoins les analyses illustrent la prédominance des femmes et de l'usage très important des plantes.

Tableau 3.1 : Répartition des tradipraticiens selon les techniques de soins déclarées

Techniques de soins	Nombre de Praticiens utilisant la technique
Observation physique	67
Observation physique + vente des plantes	57
Observation physique + divination par les esprits	12
Observation physique + divination par les esprits +vente des plantes	1
Observation physique + divination par les cauris	24

Observation physique + divination par les cauris + vente des plantes	4
Observation physique + marabout	8
divination par les cauris	1
vente des plantes	39
Observation physique + divination par le sable	6
Observation physique + divination par oracles	1
Observation physique + divination par sable, papier, cauris	1
Observation physique + divination par bâton et paume	1
Observation physique + rebouteux	1
Apprentis	1
Sans réponse	39
Total	263

Source : Sawadogo, N., Enquête PhD 2010-2011

On observe en outre que plus de la moitié des enquêtés sont des femmes. Comme, socialement, elles ont été exclues de certains rôles, leurs connaissances concernent beaucoup surtout les plantes médicinales, ainsi que le montre le tableau 2, et les bonnes pratiques de maternage.

Changements Climatiques et crise de la biodiversité au Burkina Faso

Ce qui est important pour notre propos, ce sont les implications des changements climatiques sur ces activités des femmes qui sont dépendantes de la nature. Le Burkina Faso, comme d'autres pays d'Afrique Sub-Saharienne, connaît des changements climatiques importants. Quel que soit le modèle utilisé, la variabilité et le changement climatique sont réels au Burkina Faso, avec de forts impacts sur les secteurs clés de l'économie tels que l'agriculture, les ressources en eau, l'élevage et le foresterie (CUICN 2011). Selon Bathiébo (2014), on observe une légère diminution de la pluviométrie moyenne (une baisse de l'ordre de 150 mm pourrait être observée en 2025), une légère augmentation des températures moyennes dans plusieurs centres urbains de près de 1°C avec une tendance à la hausse relative estimée à 2,5°C pour l'ensemble du pays en 2025 ; il en résulte une élévation relative de l'évaporation, atteignant la valeur de 2 000 mm en 2007, contre 1 966 mm dix ans auparavant. Malgré la relative préservation des biodiversités dans les zones rurales (Ouédraogo et al. 2013 ; Sourabié et al. 2013) force est de constater que leur dégradation est réelle (CUICN 2011 ; Bancé et al. 2011). En plus des effets des changements climatiques, les pertes des habitats de la biodiversité du fait des activités agricoles ou pastorales menacent sa préservation (Thiombiano et al. 2011). Ainsi, la biodiversité forestière du pays connait une dégradation accélérée depuis 1992. On enregistre 43 pour cent de zones dégradées en 2000 au Burkina Faso, et 60 pour cent des aires classées sont touchées à divers degrés par l'activité agricole, qui détruit chaque année 110 hectares de formations végétales. Quant à la faune aussi bien aquatique que fourragère elle est également menacée par la dégradation de ses habitats (Bélemsogbo 2011 ; Ouédraogo et al. 2011). Nombre de ses espèces animales, végétale et minérales sont utilisées dans la pharmacopée traditionnelle, qui occupe nombreuses femmes dans les villes du Burkina Faso particulièrement celle de Ouagadougou (Bougnounou 2011).

Crise de la biodiversité et menaces des opportunités informelles des femmes

Il est maintenant possible de faire le parallèle, et d'apprécier les implications de la désertification sur les opportunités des femmes qui tirent leurs revenus de l'approvisionnement de services médicaux employant des ressources végétales et ou animales.

Les conséquences des changements climatiques sur les opportunités économiques des praticiennes de la médecine et de la pharmacopée traditionnelle à Ouagadougou peuvent être considérées du point de vue écologique. La raréfaction des espèces naturelles médicinales influence directement la composition des métiers de la médecine traditionnelle à Ouagadougou. Nombre de femmes (âgées ou jeunes) avaient trouvé dans l'approvisionnement des soins infantiles et maternels traditionnels un moyen de s'intégrer à la vie urbaine, dans un contexte de chômage et de crise de la famille. Comme ces activités étaient une recomposition directe de leur rôle maternel, non seulement les hommes étaient structurellement exclus du marché – ce qui était bien pour les femmes –, mais ce marché, n'étant pas assez lucratif, décourageait les hommes.

La « dominance » des praticiennes est liée aux ressources discriminatoires dont elles disposent pour répondre à un besoin qu'un exode rural ne cesse d'alimenter, parallèlement à la déficience du système de santé moderne.

De nos jours, les conditions ont significativement changé. La raréfaction des ressources naturelles conduit à un investissement important pour se procurer de la matière première et cela joue sur le coût des soins proposés. Dans ses conditions, seules quelques femmes bien établies peuvent continuer dans le métier en se faisant des intermédiaires dans les villages qui les ravitaillent en matière première.

Cependant, l'activité est devenue une aubaine pour les hommes (tradipraticiens), car ils peuvent se déplacer très facilement, étant donné qu'ils sont culturellement favorisés et ont des moyens de transport personnels plus sécurisés. En outre, le fait que l'activité soit devenue lucrative encourage les tradi-praticiens à s'y investir.

Il émerge donc une sorte de compétition où les hommes ont maintenant tendance à envahir le marché, remplaçant les femmes qui ne peuvent plus exercer cette activité en raison des conditions d'accès à la matière première. Ces femmes, qui avaient la charge économique de leur famille, sont ainsi privées des revenus tirés de cette activité.

Il est intéressant d'explorer les transformations des rapports de genre et la situation de la femme dans ce contexte. En d'autres termes, même là où originellement elles avaient le monopole, les femmes peuvent du jour au lendemain le perdre ou ne pas le contrôler et ce en faveur des hommes. D'autre part, l'idée est que ce marché des produits de pharmacopée vendus par les femmes, a pour demandeurs d'autres femmes, puisque généralement ces activités concernent la santé de la reproduction féminine et la santé infantile. La structure sociale des consommatrices n'est pas homogène, mais la tendance est le nombre élevé des patientes provenant des couches relativement non favorisées de la ville. L'existence de cette catégorie de consommatrices forme la base sociale de cette activité ; sans cette demande une telle activité ne saurait exister.

Lorsque les produits étaient abordables, c'était une alternative aux services modernes auxquels ces patientes n'avaient pas accès. De plus les tradi-praticiennes offrent souvent des services qui n'existent même pas dans le système formel, étant donné les conditions historiques d'émergence et de développement de ce dernier. Ce marché permettait particulièrement d'améliorer relativement l'alimentation et la santé des enfants. Si les produits deviennent inaccessibles financièrement, cela risque de constituer un obstacle à l'amélioration de la santé de la mère et de l'enfant en particulier.

Il apparait donc que le changement climatique en rendant les services de la médecine et de la pharmacopée moins accessibles, rend non seulement les praticiennes encore plus pauvres, mais rend aussi la catégorie sociale des patientes, que sont les autres femmes et leurs enfants, encore plus vulnérable.

Conclusion

Le propos avancé par cet article était que les modèles d'urbanisation des villes africaines offrent des opportunités de compréhension de processus urbains plus larges, en ce sens qu'ils permettent d'analyser des processus structurels tels que la relative reproduction sociale de la position vulnérable des femmes.

Il ressort des analyses que les changements climatiques contribuent à précariser les conditions de vie de couches sociales urbaines déjà vulnérables.

Les tradipraticiennes de santé vivaient et entretenaient leurs familles avec les revenus tirés de cette activité. Au même moment, la rentabilité des soins jusqu'alors majoritairement prodigués par des femmes conduit un nombre croissant de praticiens à investir le secteur au détriment des femmes.

Références

Abbott, A., 1988, *The System of Professions : an Essay on the Division of Expert Labor,* Chicago, University of Chicago Press.

Audouin, J. & R. Deniel, 1978, *L'islam en Haute-Volta à l'époque coloniale,* Paris, L'Harmattan.

Badini, A., 1970, « *Les éléments de la personne humaine chez les Mòsé* », Bulletin de l'IFAN, vol. 41, n° 4, p. 786-818.

Bado, J.-P., 2006, *Les débuts de la médecine moderne en Afrique de l'Ouest francophone, Les conquêtes de la médecine moderne en Afrique,* Paris, Karthala.

Bado, J-P, 1997, « La santé et la politique en AOF à l'heure des indépendances », in Becker, C., S. Mbaye & I. Thioub (eds), *AOF, Réalités et Héritages,* Dakar, Direction des Archives Nationales du Sénégal, p. 1 242-1 259.

Bancé, S.B., 2011, « Mise en œuvre de la politique nationale de conservation de la diversité biologique et principales conclusions de la 10e conférence des parties », in *CUICN, Forum national sur la diversité biologique au Burkina Faso. Synthèse des résultats,* du 13 au 15 avril 2011, p. 7.

Becker, C., S. Mbaye & I. Thioub, 1997, *AOF: réalités et héritages. Sociétés ouest-africaines et ordre colonial, 1895-1960,* Dakar, Direction des Archives du Sénégal.

Belemsogbo, U., 2011, « Dynamique et contribution de la faune terrestre et de sa biodiversité à l'amélioration des conditions de vie des populations locales », in *CUICN, Forum national sur la diversité biologique au Burkina Faso. Synthèse des résultats,* du 13 au 15 avril 2011, p. 10.

Blumer, H., *Symbolic interactionism : perspective and method,* Englewood Cliffs, N. J., Prentice-Hall.

Bognounou, O. & S. Guinko, 2005, Ethnobotanique-médecine traditionnelle, Ouagadougou

Bognounou, O., « Evolution de la diversité floristique », in *CUICN, Forum national sur la diversité biologique au Burkina Faso. Synthèse des résultats,* du 13 au 15 avril 2011, p. 15.

Bonnet, D., 1994, « L'éternel retour ou le destin singulier de l'enfant », L'Homme, vol. 34, n° 131, p. 93-110.

Bonnet, D., 1988, *Corps biologique, corps social. Procréation et maladies de l'enfant en pays Mossi,* Paris, Orstom.

Brasseur, G., 1997, « Un regard géographique sur l'AOF de 1895 » in Becker, C., S. Mbaye & I. Thioub, *AOF : réalités et héritages. Sociétés ouest-africaines et ordre colonial, 1895-1960,* Dakar, Direction des Archives du Sénégal, p. 36-49.

Cassirer, E., 1972, *La Philosophie des formes symboliques,* t. II, *La Pensée mythique,* Paris, Les Éditions de Minuit.

Crowder, M., 1968, *West Africa under colonial rule,* Hutchinson, London.

Dim Delobsom, A. A., 1932, *L'empire du Mogho-Naba*, Paris, Domat-Montchrestien.

Englebert, P., 1996, *Burkina Faso. Unsteady statehood in West Africa*, Colorado, Westview Press.

Erny, P., 1968, *L'enfant dans la pensée traditionnelle de l'Afrique Noire*, Paris, Le Livre Africain.

Erny, P., 1999, *Les premiers pas dans la vie de l'enfant d'Afrique Noire*, Paris, L'Harmattan.

Fortes, M., 1975, *Œdipe et Job dans les religions africaines*, Paris: Mame-Repères.

Fortes, M., 1959, *Œdipus and Job in West African religion*, Cambridge, Cambridge University Press.

Halpougdou, M., 1992, *Approche du peuplement pre-dagomba du Burkina Faso : Le Yônyônse et les Nînsi du Wubr-Tênga*, Stuttgart, Franz Steiner Verlag.

Hughes, E. C., 1993, *The sociological eye : selected papers*, New Brunswick, Transaction Books.

INSD, 2009, *Recensement General de Population et de l'Habitat 2006. Ouagadougou* Ouagadougou, INSD.

Ki-Zerbo, J., 1978, *Histoire de l'Afrique Noire*, Paris, Hatier.

Kouanda, A., 1996, « La progression de l'islam au Burkina Faso pendant la période coloniale, » in Massa, G. & Y. G. Madiega, *La Haute-Volta coloniale*, Paris, Karthala p. 233-248

Lallemand, S., 1971 « La relation mère-enfant en milieu mossi traditionnel », Rapport de mission, Documents voltaïques, notes, vol. 8-5, n° 1.

Latour, B., 1987, *Science in action : How to follow Scientists and Engineers through Society* Cambridge, Mass, Harvard University Press.

Massa, G. & Y. G. Madiega, 1996, *La Haute-Volta coloniale*, Paris : Karthala

Monné, R., 1994, *Contribution à l'Étude Juridique du Droit à la Sante en Afrique : Réflexion à partir de l'exemple du Burkina, thèse en droit public, Université de Bordeaux I.*

Elias, N., 1969, *What is sociology ?* London, Hutchinson.

Ouédraogo, A., A. M. Lykke, B. Lankoandé & G. Korbéogo, 2013, « Potentials for promoting oil products identified from traditional knowledge of native trees in Burkina Faso », *Ethnobotany Research & Applications*, vol. 11, p. 71-83.

Ouédraogo, R. L., N. D. Coulibali & F.-Ch Ouedraogo, 2011, « Dynamique actuelle et contribution de la biodiversité aquatique à la satisfaction des besoins des populations au Burkina Faso », in CUICN, *Forum national sur la diversité biologique au Burkina Faso. Synthèse des résultats*, du 13 au 15 avril 2011, p. 11.

Parsons, T., 1991, *The social system*, London, Routledge.

Poulet, E., 1970, *Contribution à l'étude des composantes de la personne humaine chez les mossi*, doctorat de 3ᵉ cycle de philosophie, Université de Poitiers.

Rabin, J., 1979, *L'enfant du lignage. Du sevrage a la classe d'âge*, Paris, Payot.

Robert Park, R. E., 1936, « Human ecology », in Lin, J. & C. Mele, *The urban sociology reader*, Abingdon, Routledge, p. 83-90.

Sakanlé, M., 1960, *Souveraineté nationale et problèmes sanitaires internationaux*, Dakar 24 février.

Sawadogo, N., 2006, *Problématisation de la maladie de l'enfant et concurrences dans l'espace thérapeutique à Lougsi*, mémoire de DEA, UCAD, Dakar.

Sawadogo, N., 2013, *Professions and the public interest*, PhD Thesis, University of Nottingham.

Sedogo, V. 2006, « Bref aperçu des activités sociales, économiques et culturelles », in Hien, C.P. & M. Compaore (dir.), *Histoire de Ouagadougou des origines à nos jours ;* 2ᵉ édition, Ouagadougou, DIST, p. 99-126.

Simporé, L., 2009, « Le geste de Wubri, le yagenga-faagda ou neveu liberateur », in Hien, C. & M. Gomgnimbou, (dir.), *Histoire des royaumes et chefferies au Burkina Faso précolonial,* Ouagadougou : DIST, p. 159-203.

Simporé, L. & D. Nacanabo, 2006, « La mise en place du peuplement et des institutions politiques » in Hien, C. P. & M. Compaoré (dir.), *Histoire de Ouagadougou des origines à nos jours,* 2ᵉ édition, Ouagadougou, DIST, p. 27-67.

Skinner, E. P., 1989, *The Mossi of Burkina Faso : Chiefs, Politicians, and Soldier,* Prospect Heights, Waveland Press.

Sourabié, T.S., N. Some, O. Bognonou, Y. Ouattara & J.-B. Ouedraogo, 2013, « Ethnobotanical and Ethnopharmacognostical Survey on Medicinal Plants of Malon Village And Surrounding In The Cascades Region (Burkina Faso) », *Iosr Journal Of Pharmacy,* V. 3, n° 2, p. 11-15.

Suret-Canal, J., 1977, *Afrique Noire. L'ère coloniale 1900-1945*, Paris, Éditions Sociales.

Thiombiano, A., A. Doulkoum, R. Diebre & C. Honadia, 2011, « Importance, état et dynamique de la biodiversité forestière dans un contexte de variabilité et de changement climatique au Burkina Faso », in CUICN, *Forum national sur la diversité biologique au Burkina Faso. Synthèse des résultats,* du 13 au 15 avril 2011, p. 9.

4

Saint-Louis du Sénégal, les « aventuriers » de la terre

Adrien Coly et Fatimatou Sall

Introduction

Saint-Louis du Sénégal est une ville localisée au niveau de l'estuaire du fleuve Sénégal. Elle fut capitale à l'époque coloniale et bénéficie ainsi d'un long processus d'urbanisation. Elle est un laboratoire des pratiques urbaines des villes inscrites sur des sites d'eau peu étendus et peu favorables à l'extension spatiale.

L'urbanisation engendre des dynamiques territoriales résultant d'un ensemble de processus physiques, biologiques et anthropiques interconnectés. Les particularités géographiques de la ville de Saint-Louis font que le territoire urbain est une réalité complexe qui ne peut s'interpréter qu'à travers une analyse systémique (Richard 1975). La considération de la ville de Saint-Louis en tant que Système socio-écologique (SSE) oblige à la prise en compte de variables écologiques, de variables humaines et de variables à la fois écologiques et humaines pour rendre compte des faits de vulnérabilité.

L'analyse de la vulnérabilité urbaine dans cet article, tente d'en recentrer les causes par leur mise en relation – pour expliquer les faits qui s'observent sur le territoire urbain – et d'en établir la chaîne de causalité. Cette démarche s'inspire de l'approche de modélisation par les diagrammes causaux, qui permet la détermination des réseaux causaux et une formalisation des liens causaux de la territorialisation urbaine à Saint-Louis, sur la base d'une entrée sociale, physique, et spatiale, mettant en lumière le niveau de vulnérabilité territoriale.

Le territoire est le résultat des processus physiques sociétaux et spatiaux qui s'opérationnalisent à travers le système d'utilisation et la situation des ressources foncières. Dans le contexte d'une ville vulnérable comme Saint-Louis, la variable aléa est multiple et regroupe les phénomènes liés à la mer, au fleuve, et à la pluie. L'eau est au cœur des processus par sa présence ou son absence, et devient l'hypothèse principale à l'édification de l'habitat.

Dans les grandes lignes, la vulnérabilité à Saint-Louis combine les conditions d'ancienneté de l'habitat (village/lotissement), la forme de l'habitat (ancien/nouveau) et le type d'habitat. Ces critères permettent de se donner une représentation extérieure des conditions du social ainsi que des possibilités existantes de faire face aux aléas.

La variable spatiale quant à elle apporte un niveau d'affinement à la caractérisation du risque par l'exposition que les populations ou les actifs ont par rapport à un aléa. Ainsi, les zones localisées assez loin de la mer ne sont pas concernées par le phénomène, alors qu'elles peuvent être soumises à l'inondation au même titre que les sites plus éloignés. La topographie peut également différencier le niveau d'exposition, de même que les aménagements et la typologie des technologies utilisées pour faire face aux aléas.

Ainsi, de prime abord, certains quartiers de Saint-Louis sont désignés comme des *hotspots* dans le contexte des problèmes à risques. Les aventuriers de la terre participent en ces lieux aux processus territoriaux. De façon consciente ou pas l'aventurier de la terre se localise dans un espace exposé à un aléa.

Cette communication pose un regard sur les pratiques d'appropriation de l'espace qui s'observent dans les zones humides et s'interroge sur leurs impacts sur la croissance de l'habitat à Saint-Louis et leur implication dans la vulnérabilité urbaine.

Saint-Louis du Sénégal, un processus urbain lié à la dynamique des zones humides

La ville s'est développée sur un territoire dominé par des infrastructures bleues et vertes. Le bâti se structure autour de ces différentes unités écologiques donnant à la cité estuarienne une morphologie assez particulière sous forme d'archipel (Figure 1).

Figure 4.1 : Saint-Louis du Sénégal (Cluva 2014)

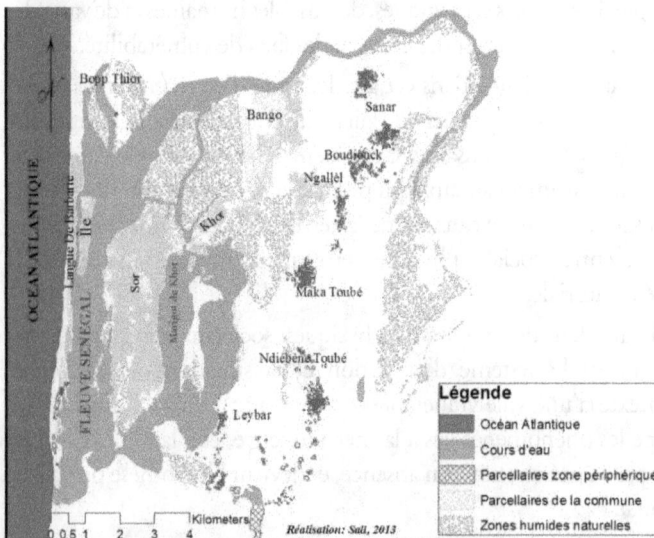

le morcellement du territoire est le fruit de processus liés à l'évolution des écosystèmes humides et aux stratégies d'aménagement.

Un contexte marqué par les changements climatiques

Les années 1960-1970 ont été marquées par une sécheresse persistante dans tout le Sahel avec une aridité chronique sur trois décennies (Sy 2008 d'après Sagna 2000). Cet avènement climatique n'a pas épargné la cité estuarienne.

Figure 4.2 : Évolution de la pluviométrie à Saint-Louis par rapport aux moyennes sèche et humide et à la normale

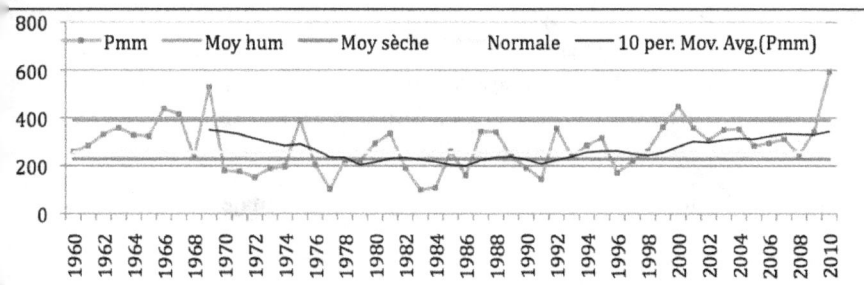

L'analyse de l'évolution de ce paramètre indique une baisse de la moyenne décennale à partir de 1976 en deçà de la normale homologuée (256 mm) et même de la moyenne sèche à partir de 1978. La moyenne humide n'est jamais atteinte, pour cet intervalle allant de 1960 à 2010, par la moyenne mobile, excepté pour des années singulières concernant lesquelles des pics ont été enregistrés pour la station de Saint-Louis. Toutefois, une tendance à l'humidité est observée à nouveau pour les années 2000, avec des moyennes décennales mobiles supérieures à la normale. L'année 1999 marque la période où on note une régularité de la pluviométrie à une moyenne annuelle supérieure à la normale, excepté pour 2008.

La baisse soutenue de la pluviométrie en concomitance avec une hausse éminente de son évapotranspiration (Sagna 2000) a conduit à un indice d'aridité constamment élevé (0,27) pour la ville de Saint-Louis (Coly et al. 2013), avec beaucoup d'années déficitaires entre 1969 et 1998.

Figure 4.3 : Analyse des années sèches et humides de 1954 à 2012

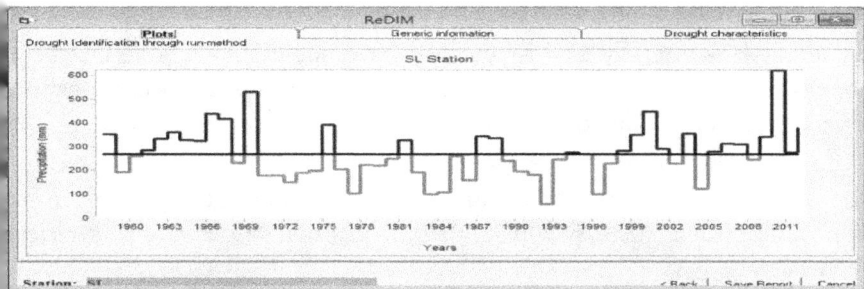

Une régression des systèmes écologiques amorcée durant la sécheresse

Cette sécheresse climatique a fortement influé sur le bilan hydrologique à traver tout le bassin du fleuve Sénégal, et notamment dans l'estuaire, avec une baisse des écoulements de près de 40 pour cent (Gac & Kane 1985). Ceci est à l'origine d'une exondation des plans d'eau évaluée à 3,64 pour cent sur Saint-Louis et son arrière-pays entre 1972 et 2003 (Sall 2012 d'après Diakhaté & Diallo 2007).

Un accueil de populations sur les espaces exondés

Dans les années 1960-1970, avec la grande sécheresse du Sahel, qui a impulse l'exode rural, Saint-Louis a connu une croissance démographique fulgurante avec un taux de 2,4 pour cent qui s'est maintenu dans les années 1980-1990 du fait du retour massif de populations suite au conflit sénégalo-mauritanien en 1988, et de l'ouverture de l'université Gaston Berger de Saint-Louis en 1990 qu a engendré de nouvelles dynamiques urbaines.

Figure 4.4 : Évolution de la population et événements majeurs

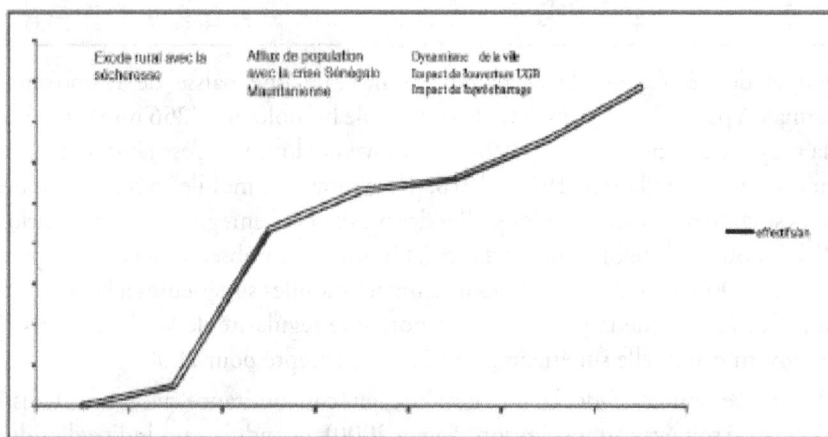

Source : Coly et al. Ss presse

La situation démographique a amorcé l'extension du tissu urbain à l'assaut d'espace exondé durant la sécheresse. L'assèchement des zones écologiques humides a favorisé la propagation de la ville.

Simple faubourg, Sor s'est agrandi, au fur et mesure de la croissance de la population, et suite à la saturation de l'île et de la Langue de Barbarie, par des séries de créations de quartiers nouveaux.

Les sites d'expansion sont de manière générale des zones d'anciens marécages, de vasières, de cuvettes de décantation, de terrasses marines et chenaux de marée de marécage (Kane 2003, Sy 2009 ; 2010). L'extension s'est faite vers les anciens dépôts de vase.

Figure 4.5 : L'évolution du bâti à Saint-Louis entre 1973 et 2009

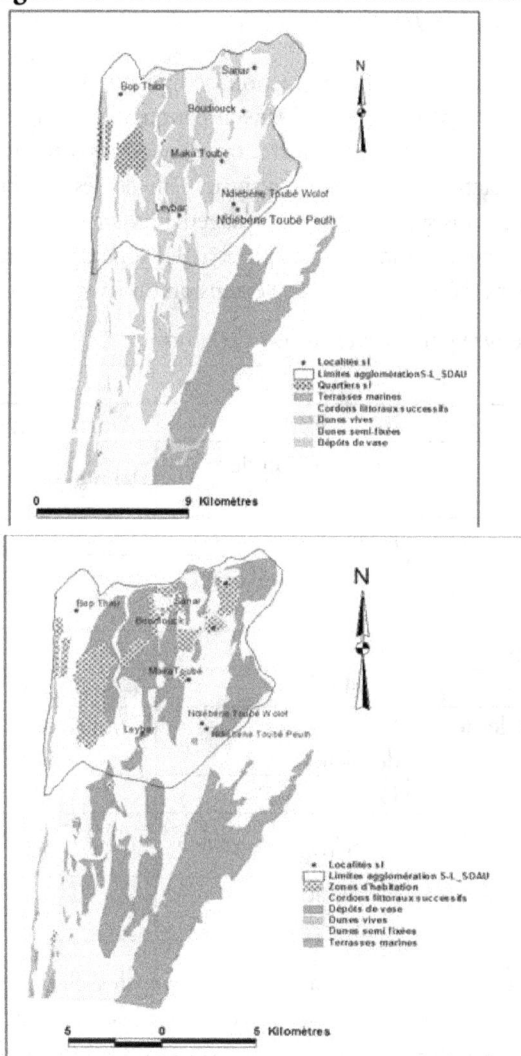

Source : Sall 2011

Les aventuriers de la terre : le processus de production du territoire d'un site sous contraintes

Le processus de construction du territoire urbain de Saint-Louis est dynamisé par le besoin en logement, qui s'explique par la pression démographique qui s'exerce sur un site dont l'espace est limité par la présence de l'eau. Sur les espaces marginalisés, les populations migrantes (désignées sous le vocable de « *doli ndar* ») ou les autochtones (appelées « *domou ndar* ») ont mis en œuvre différents processus d'appropriation de la terre.

Des formes d'appropriations de l'espace s'articulant autour d'axes structurants et soutenues par des actions de remblai

L'installation des populations dans les zones humides non aedificandi s'est faite progressivement à la faveur du recul des écosystèmes. Dans la conquête de l'espace libéré par l'eau, les aventuriers de la terre ont mis en œuvre trois catégories d'actions (Tableau 1). Il s'agit des actions réactives, des actions d'ajustement et des actions anticipatives. Les actions réactives consistent en des aménagements structurants, les actions anticipatives et d'ajustement confortent et appuient les aménagements structurants. Ces derniers continuent d'être des socles de l'extension du tissu urbain.

Tableau 4.1: Actions mises en œuvre par les aventuriers de la terre

Zones	Risques	Actions		
		Anticipative	Réactive	Ajustement
Littoral	Érosion		Construction de murettes de protection	Reboisement
			Batteries de pneus	
	Inondation	Remblai déchets	Remblai de sable ou gravats	
		Digues de déchets		
Bas fonds	Inondation	Remblai	Remblai	Reboisement
		Vidange des fosses septiques	Pompage des domiciles	Remblai de sable ou gravats
		Digue de sable	Digues route	
		Surélévation des maisons	Déménagement	
		Ouverture de la brèche	Surélévation du mobilier	

Il faut considérer la route, le tas de déchets et la digue comme des éléments essentiels à l'occupation des zones exondées. Dans les quartiers, le bitumage d'un axe de communication avec les normes de construction requises amène à une surélévation de l'infrastructure pour garantir sa sécurité. Cet aménagement devient une opportunité, pour les propriétaires de parcelles, de se lancer dans un effort de remblaiement qui s'arc-boute sur la route ; le développement du bâti se fait ainsi de la route vers la zone la plus basse.

La route nationale constitue un exemple d'une charpente du développement de la suburbanisation. À la suite de la création de l'université, l'axe allant vers l'université Gaston Berger devient l'espace qui voit se dédoubler la ville (Sarr 2002), suite aux interventions sporadiques de la municipalité avec quelques équipements primaires : bornes fontaines, postes de santé, marchés… (Wade & Diop 2000) mais surtout grâce à l'auto-construction, fait de particuliers.

La mise en place de la route et des digues de protection a donné aux quartiers de Darou et Guinaw rail une possibilité de s'étendre dans la vasière de Saint-Louis. La mise en place de cette route a contribué à la reconfiguration du réseau hydrographique. La hauteur de cette infrastructure avait amorcé l'exondation des terres, créant des espaces retranchés des zones de crues.

Deux stratégies concomitantes ont été mises en œuvre pour conquérir l'espace. L'une réplique la démarche faite au niveau des routes, l'autre amorce l'effort de remblai par les déchets.

Les formations de petit noyau d'habitat au niveau des promontoires se sont rapidement densifiées avec la mise en place des routes et des digues qui constituent des charpentes de la suburbanisation.

Figure 4.6 : Schémas de conquête de l'espace

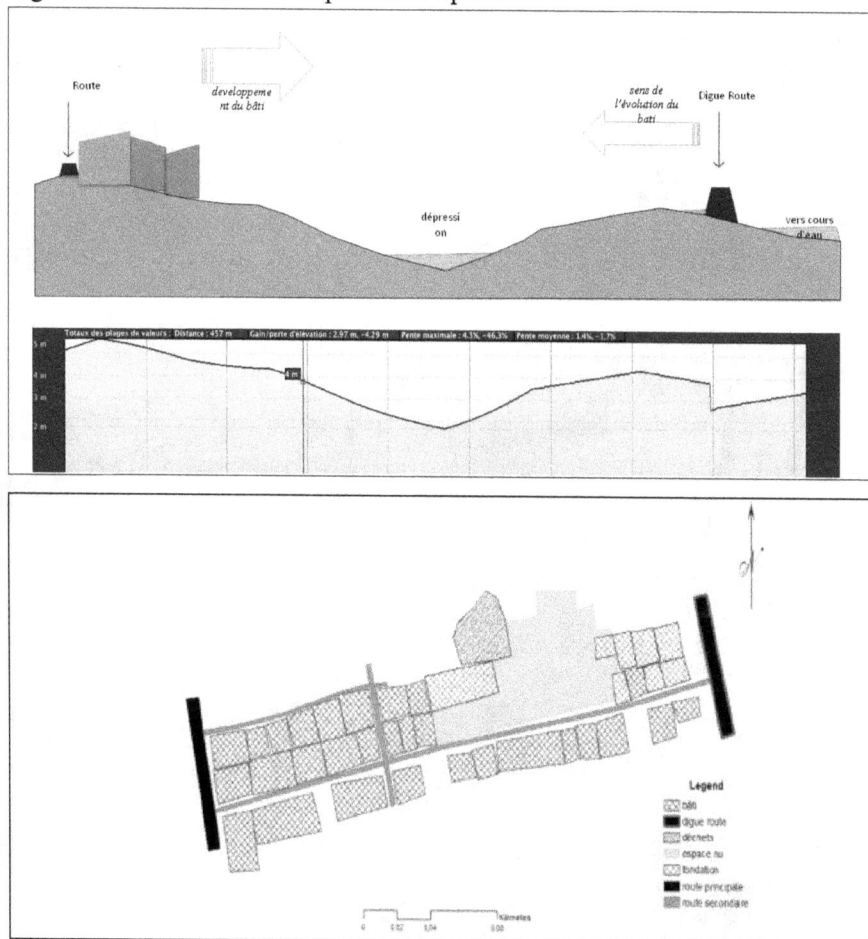

Ainsi les deux quartiers ont pu s'étendre et dans la zone étudiée, les superficies ont augmenté respectivement de 54 pour cent et de 80 pour cent (**Figure 7**).

Figure 4.7 : Méthodes de conquête des bas-fonds

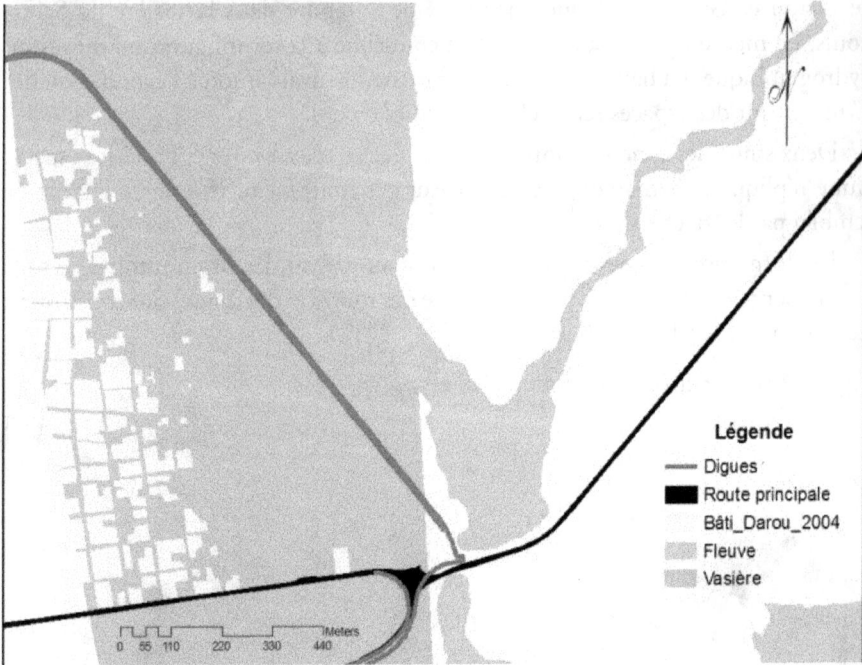

a

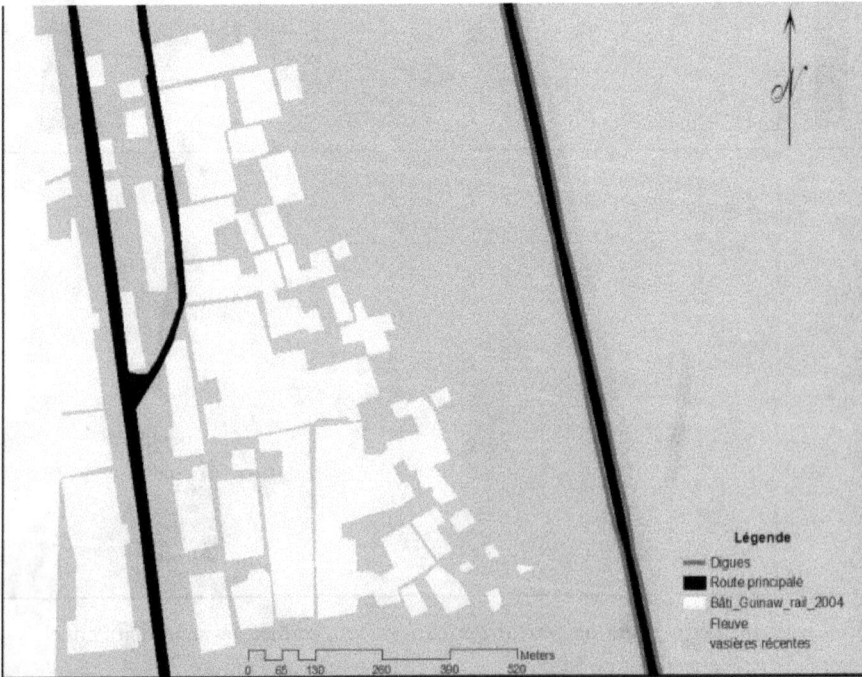

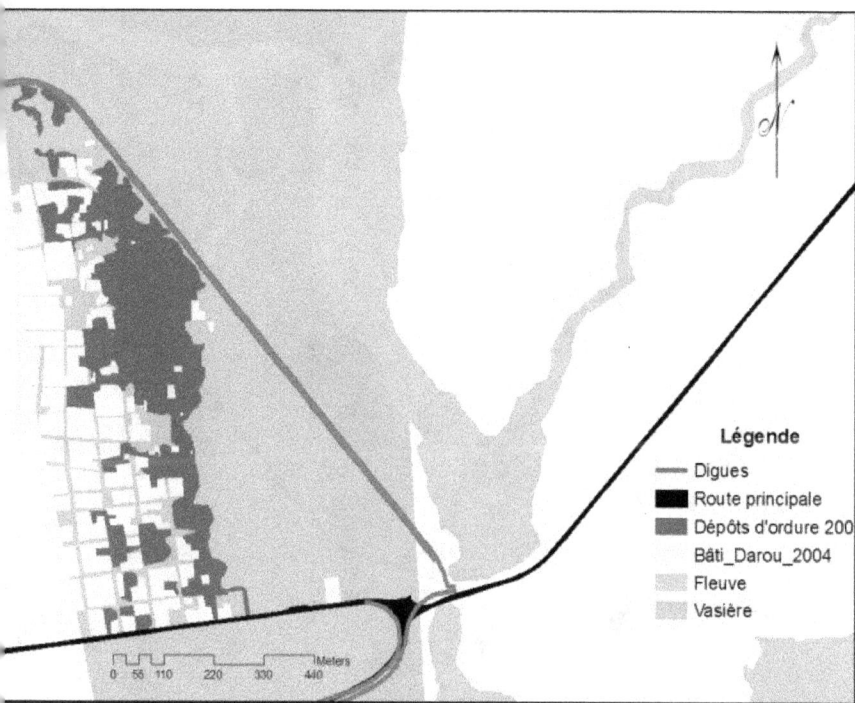

Légende
Digues
Route principale
Dépôts d'ordure 200
Bâti_Darou_2004
Fleuve
Vasière

0 55 110   220   330   440   Meters

Légende
Digues
Route principale
Dépôts Guinaw rail2004
Bâti_Guinaw_rail_2004
Fleuve
vasières récentes

0   65   130   260   390   520   Meters

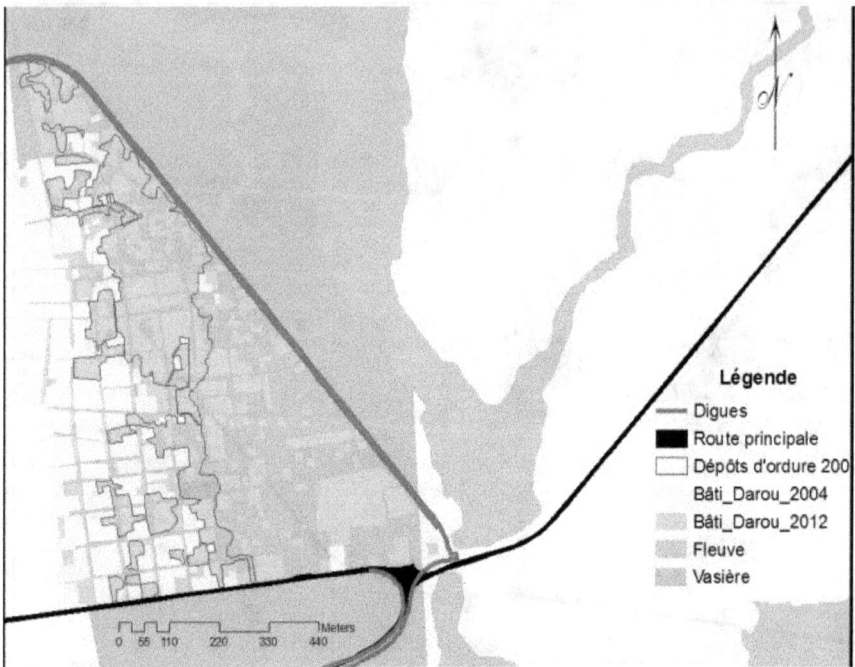

Légende

— Digues
■ Route principale
□ Dépôts d'ordure 200
Bâti_Darou_2004
Bâti_Darou_2012
Fleuve
Vasière

0   55 110   220   330   440 Meters

c

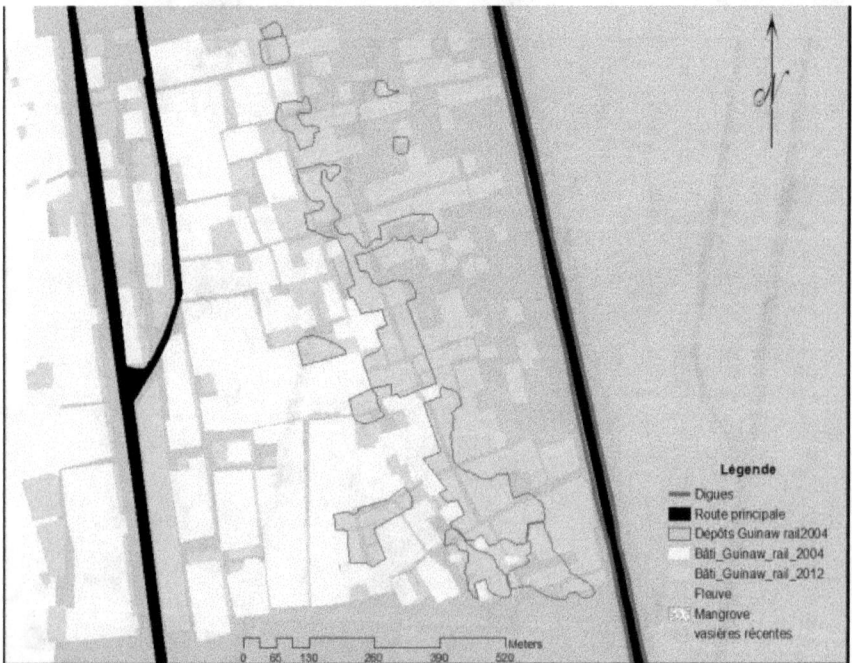

Légende

— Digues
■ Route principale
□ Dépôts Guinaw rail2004
Bâti_Guinaw_rail_2004
Bâti_Guinaw_rail_2012
Fleuve
Mangrove
vasières récentes

0   65 130   260   390   520 Meters

a) situation en 2004, b) remblais par dépôt d'ordure (en gris), c) extension urbaine
survenue en 2012 sur les zones remblayées

Saint-Louis, des écosystèmes chevauchés par l'urbain

La reconfiguration du réseau hydrographique crée des espaces d'accueil pour les aventuriers de la ville et l'urbanisation en est devenue très vite incontrôlable. La mise en place de cette route pour protéger les populations déguerpies de l'île et les sinistrés des raz de marée de la Langue de Barbarie, relogés dans Sor, entraîne une seconde vague d'urbanisation.

On assiste à des installations spontanées avec des remblais par ordures donnant naissance à un enchevêtrement de maisons et de ruelles tortueuses.

Le résultat d'un tel processus a pour conséquence une urbanisation diffuse, marquée par la formation de pôles urbains, et une série d'extensions du bâti au-delà des limites communales (Wade & Diop 2000).

Cette situation traduit un double paradoxe du caractère urbain de Saint-Louis. Malgré un taux d'urbanisation de 10 pour cent par an entre 1970 et 1988, les infrastructures grises, c'est-à-dire les espaces aménagés et construits, ne représentent que 15 pour cent de l'ensemble du tissu urbain. Les espaces naturels et semi-naturels (infrastructures vertes et en infrastructures bleues) s'étendent sur 85 pour cent du territoire urbain. On peut ainsi noter une progression du bâti qui gagne sur les autres espaces (Figure 8).

Figure 4.8 : Évolution de l'occupation du sol entre 2003 et 2011 en ha

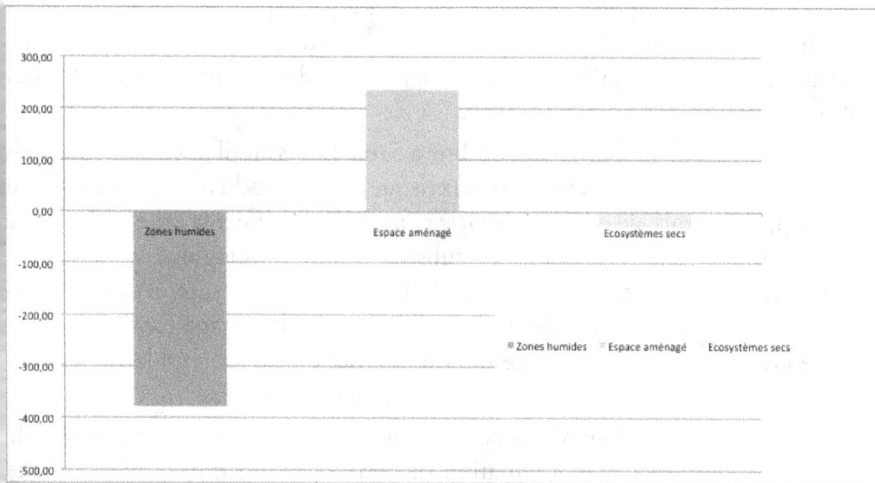

L'urbanité source de la vulnérabilité urbaine à Saint-Louis ?

L'urbanisation à Saint-Louis a pour résultante la création d'une chaîne de vulnérabilité pour les populations à la recherche de la « terre promise ». En effet ces dernières, qui se sont propulsées en ville suite aux contraintes du milieu rural durant la sécheresse, se voient piégées de nouveau en ville par le climat (Figure 9).

Même si la sécheresse a continué au courant des années 1990, avec un indice de 0,1 (Sy 2008 d'après Sagna et al.), la modification de la nature des pluies, avec l'importance croissante des averses, a fini par installer les populations dans une vulnérabilité structurelle, même si au plan climatique, les pluies enregistrées ne constituent pas toujours des pluies exceptionnelles.

Figure 4.9 : Urbanisation et vulnérabilité à Saint-Louis

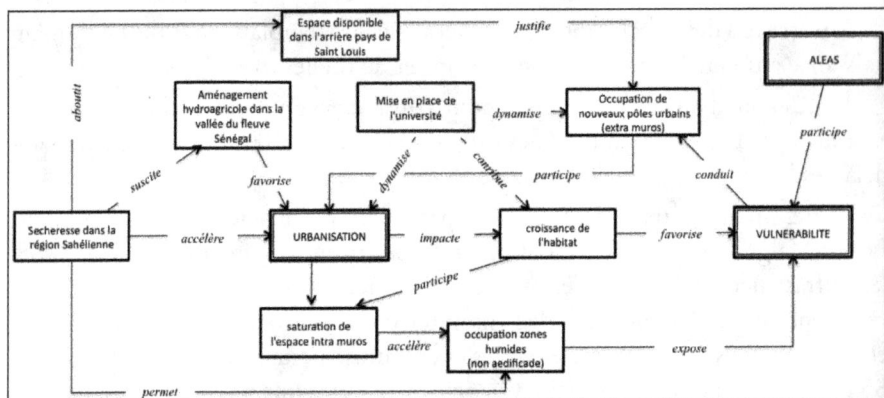

Source : Coly et al. ss presse

Les processus de formation des établissements humains laissent transparaître « le reflet d'un nouveau rapport de l'homme à son espace que des années de faibles pluviométries ont contribué à favoriser, sans qu'aucune politique importante de viabilisation de ces milieux n'accompagne cette dynamique » (Sène & Ozer 2002).

Le schéma urbain, tel qu'il s'est décliné à Saint-Louis au fil des ans, n'a pas pris en compte les aléas pouvant concourir aux risques. L'analyse du risque d'inondation montre qu'il y a trois aléas qui peuvent y concourir : la pluie, la crue, et la nappe avec, comme facteur favorisant, le substrat géomorphologique et le réseau d'assainissement (Coly et al. 2011, Sall 2013). Or dans les mesures d'urbanisation prises depuis l'époque coloniale, les énergies se sont davantage focalisées sur la variable crue avec la mise en place des digues, des quais, les remblais… Ce qui actuellement ne préserve pas les populations de ce phénomène.

Les populations confrontées à ce problème, que l'on appelle aventuriers de la terre, font montre d'une acceptation de cette situation qui, fondée sur une certaine idée de leur capacité à y faire face avec des aménagements technologiques plus ou moins appropriés et une stratégie de mobilité saisonnière pour certaines populations (5 451 en 2010 selon la roix Rouge), indique de fait une résignation très forte de leur part.

Si cette population a tendance à être identifiée aux originaires du monde rural, tant qu'il y a une cohabitation hybride une discrimination s'avère difficile à faire au plan spatial car, en réalité, puisque les acteurs proviennent de toutes les couches (autochtones ou étrangers, chômeurs ou travailleurs, riches ou pauvres…), cela se reflète dans l'image que donnent les paysages urbains à Saint-Louis (Lapierre & Luchetta 2003).

L'extension du bâti sur des zones *non aedificandi* occupées chaque année par la pluie, fait penser que ces saint-louisiens en sont venus à décider de vivre avec l'eau et que c'est l'attitude qui sera adoptée dans la perspective des changements climatiques (Saint-Louis, horizon 2030, PGIZC – Saint-Louis). Ce choix de produire la ville positionne Saint-Louis sur l'échiquier des villes résilientes mais semble être source de vulnérabilité urbaine.

Conclusion

La ville de Saint-Louis a été créée par des « aventuriers de la terre ». Que ce soit à l'époque coloniale où à celle d'après-sécheresse, le territoire urbain s'est structuré progressivement avec de nouveaux arrivants qui se sont établis sur des sites peu propices en déployant des trésors d'énergie et d'ingéniosité pour rendre le site habitable.

Les zones de vasières exondées durant la sécheresse de 1970 ont accueilli des populations qui, progressivement, ont gagné un espace. La recherche de logement a conforté ces stratégies, que les autorités ont accompagnées avec des programmes de structuration.

L'installation sur des sites essentiellement constitués de vasière est un des éléments déterminants de la vulnérabilité territoriale d'autant plus que la structure de la vulnérabilité socioéconomique semble diffuse au plan spatial. La ville de Saint-Louis est à l'image de ses quartiers, le résultat d'une construction qui laisse apparaître le rôle de l'aléa dans la morphologie urbaine.

Bibliographie

Richard, J. F., 1975, *Paysages, Écosystèmes, Environnement : une approche géographique*, L'espace géographique, n° 2 Tome IV, Paris ORSTOM, p. 81-92

Gac, J. Y., Kane 1985, *L'invasion marine dans la basse vallée du fleuve Sénégal*, ORSTOM, 64 p.

WADE, C. S. & O. Diop, 2000, « La croissance urbaine et ses incidences géographiques sur l'espace rural : le cas de la Commune de Saint-Louis et la Communauté rurale de Gandon », *Revue AFRISOR (Afrique-Sociétés-Recherches). Revue des Sciences Sociales et Humaines*, n° 1, p. 13-58.

Sall, F., 2013, *Le Profil de vulnérabilité, un outil d'aide à la décision pour la gestion durable des zones humides de Saint-Louis*, mémoire de master, Département environnement, université Senghor d'Alexandrie, 52 p.

Coly, A. et al., 2011, *Report on Climate Related Hazard in the Selected Cities* (Saint-Louis) D5. 2, CLUVA, 51 p.

Coly, A., F. Sall & G. Weets, 2013, *Changements climatiques et Vulnérabilité urbaine en Afrique. Saint-Louis du Sénégal*, D5. 8, CLUVA

Kane, C., 2003, *Étude diachronique des espaces habitables de la commune de Saint-Louis des origines à nos jours : Éléments de cartographie de limites complexes*, mémoire de maîtrise, section de Géographie, UGB, 85 p.

Lapierre & Luchetta, 2002, *Dynamique du risque d'inondation à Saint-Louis du Sénégal, mémoire de maîtrise*, Université Gaston Berger de Saint-Louis

Sall, F., 2011, *Changements climatiques et Impacts sur les zones humides de la ville de Saint-Louis*, mémoire de master1, section de Géographie, UGB, 90 p.

Sall, F., 2012, *Changements climatiques et Impacts sur les zones humides de la ville de Saint-Louis*, mémoire de master1, section de Géographie, UGB, 88 p.

Sarr C., 2002, « Inscription urbaine dans l'écosystème binaire saint-louisien », *Revue de Géographie de Saint-Louis,* n° 2 Janv 2003, p. 50-64.

Sène, S. & P. Ozer, 2002, « Évolution pluviométrique et relation inondations – événements pluvieux au Sénégal », Bulletin de la société géographique de Liège, 42, 2002, p. 27-33

Sy A. A., 2009, *Les dunes littorales de la Grande Côte sénégalaise. Dynamique actuelle et conséquences sur les espaces maraîchers,* mémoire de master2, section de Géographie, UGB, 193 p.

Sy B. A., 2008, *Milieux, sécheresse climatique et érosion éolienne étude géomorphologique du Sahel sénégalais,* Thèse de doctorat d'État, UGB, 429 p.

SY B. A., 2010, « Caractéristiques physiques et perspectives d'aménagement du site urbain de Saint-Louis au Sénégal », communication scientifique, le 12 avril.

SY B. A., 2010, « Géomorphologie et assainissement urbain : Exemple du Faubourg de Sor à Saint-Louis/Sénégal », communication scientifique.

5

A New Cartography of International Cooperation: Emerging Powers in Sub-Saharan Africa – The Case of Biofuels Promotion by Brazil in Senegal

M.A. Gaston Fulquet

Introduction

The relationship between energy and climate change is one of the most relevant discussions in the dawning of the twenty-first century. Traditional fossil fuels stand as one of the main causes responsible for the rising levels of greenhouse gas emissions in the atmosphere. Because of this, the introduction of alternative energy sources has become a top priority for many governments around the world. Among them, liquid biofuels were internationally supported as one of the few viable alternatives for fighting climate change in the short run.

In a context of an increasing global energy demand, a plethora of international cooperation initiatives are being fostered as a way to promote this energy alternative. Interestingly, as the dominance of the classical western powers in leading the debates and actions around development and cooperation is eroding, emerging powers have begun to play an increasingly significant role in redefining the architecture of international cooperation (IC). Through the provision of technical assistance and investment agreements, South-South cooperation (SSC) initiatives are becoming a tool by which emerging economies promote biofuel production in Africa.

Within this group, Brazil has not only become one of the global biofuel supporting countries but also a new donor of IC becoming a leading sector player in Sub-Saharan Africa (SSA) after its extensive agricultural trajectory and its renowned expertise in the production and use of biofuels. As Lechini noted 'supported by governmental and private actors, Brazilian diplomacy is having an impact on the regional and international stages, within a context in which SSC is presented as a strategy seeking to strengthen the capacities of developing countries' (Lechini 2011: 215). Nonetheless, recent studies (Richardson 2011 Franco et al. 2011; Ferreira de Lima 2012) identify that activities related to large-scale monoculture for the production of this alternative source of energy is turning to be one of the main causes of socio-environmental conflicts and disputes in Africa as it already happened elsewhere (Delgado Ramos et al 2013 Fulquet 2015). In this direction, biofuels become an arguable tool for Climate Change (CC) mitigation, currently under scrutiny also in SSA countries.

The study of how the new dynamics of South-South cooperation are affecting the political economy and sustainability of rural development and energy diversification in countries in Africa remains relatively unexplored. By taking the case study of the relationship between Brazil and Senegal, this chapter also proposes to problematise the progress and setbacks experienced by the biofuel sector in that African country, reflecting a dilemma of global reach: the absence of certainties around whether liquid biofuels constitute a sustainable energy alternative to cope with climate change. In order to do this, the chapter reviews some central concepts revolving around emerging powers, SSC and agrarian development. The analysis draws on interviews conducted both in Brazil and Senegal as a way to reflect the diverse interplay of actors, motivations, interests and tensions observed in the context of biofuel developments through international cooperation.

In light of that, our first section will introduce the contemporary phenomenon of the rise of emerging powers as new international cooperation donors through SSC actions. The subsequent section will provide a general panorama on recent developments regarding biofuel policies, actors and conflicts in SSA countries. Section 3, will be dedicated to the analysis of the broader opportunities and limits introduced by Brazilian international actors in the biofuel sector in Senegal as a case study. Finally, by assessing the interrelation between IC and sustainable development, this article concludes that biofuels introduce deep asymmetries and inequalities that reflect some sort of stratification between what Acharya (2014) calls the 'power South' and the 'poor South'.

Emerging Powers and South-South Cooperation

The understanding of recent trends in the development of liquid biofuels in SSA falls under a more general discussion on the reach and limits of a changing cartography of international cooperation worldwide. Since year 2000, there has been a redistribution or shift of economic and political capacities among states in the international system. A wide range of conceptualisations like 'unstable multipolarity' (Humphrey & Messner 2006), 'multiregional global order' (Hurrel 2007), 'multi-multipolarity' (Nolte 2008), 'growing multipolarity' (Nederveen Pieterse 2008) or 'interpolarity' (Badie 2013) have been recently coined as a way to characterise this phenomenon. All these scholars agree on the fact that the re-ordering of the global political economy is associated with the capacities of a new set of players to directly or indirectly affect the nature and reach of global interactions.

Additionally, over the past decade, new transregional coalitions have emerged. The emerging powers groupings of India, Brazil and South Africa (IBSA) and those together with China and Russia (BRICS), are clear expressions of soft balancing strategies vis-à-vis the hard core of the G8, looking in this way to shape new paths for their international political positioning in global governance. This tendency is complemented by the new role that this rising countries are playing in international development cooperation. Countries like China and Brazil have in recent years become influential actors in the former closed circle of cooperation donors by means of implementing South-South cooperation initiatives with other developing countries.

The foreign actions of many emerging countries in present times tend to increasingly incorporate South-South cooperative actions. Among the heterogeneous group of the so-called 'third world' countries, today's emerging economies stand out from the rest of the developing countries for their well-developed technical capacities. This internal advantage has been increasingly used internationally over the last few decades, promoting the exchange of technical knowledge and management with the goal of boosting institutional and human developmental capacities elsewhere (Pérez de Armiño 2000). In recent years, the volume of resources and the number of South-South partnerships and programmes have increased significantly, and technical cooperation has become a key component of IC initiatives propelled by emerging economies such as China, India and Brazil.

Taking this component into consideration, South-South cooperation has been defined[1] as the 'process by which two or more developing countries acquire individual or collective capacities through cooperative knowledge, capacity-building and technological know-how' (SEGIB 2008:16) in a wide array of policy areas such as public health, education, social development, agriculture, food security and energy, among others.

However SSC is a more comprehensive concept that includes other actions beyond technical cooperation among developing countries. In this direction, as the United Nations Trade and Development Conference observed, SSC refers to 'the processes, institutions and arrangements designed to promote political, economic and technical cooperation among developing countries in pursuit of common development goals' (UNCTAD 2010: 1).

Therefore, IC, as a complex component of foreign policy, can be better understood if organised around three different but interrelated dimensions. In the first place, it involves a political dimension that refers to a process of bilateral political dialogues for policy coordination led by governmental actors representing the cooperating countries. It does not necessary exclude other interest groups such as private sector and civil society representatives. In general terms the political dimension tends to be pushed forward by pre-existing ideas and interests with the goal of maximising certain political objectives.

In the second place, IC is also formed by a technical-scientific dimension, including all actions oriented towards the creation of stronger ties between the technical and scientific communities from both cooperating parties. This is mainly done by fostering technical collaboration between State bureaucracies, private sector representatives and cooperation between scientific institutions. It usually involves actions associated with information exchanges through technical meetings, capacity building programmes or joint research actions.

Finally, an economic dimension can also be identified. In this sense, economic cooperation seeks to foster the incremental interdependence between the collaborating parties in terms of trade and investments with the goal of diversifying exchanges of goods and services. It could lead to progressive and reciprocal trade liberalisation by implementing preferential tariff agreements between the partners.

Actions developed at a South-South level are generally supported by bilateral cooperation framework agreements, through which the core of capacity-building, knowledge and technology transfers is complemented by investments instead of direct monetary transfers (Hochstetler 2012).

Despite being guided by the principles of respect for national sovereignty, national ownership and independence, equality, non-conditionality and non-intervention in domestic affairs, it would be naïve to state that SSC is totally emptied of the strategic national interests of donor countries. As Smith stated, when comparing historical and contemporary SSC, the 'once strong sense of solidarity and unified purpose seems to have given way to more pragmatic and self-interested considerations among states in the Global South' (Smith 2014: 2). Therefore, it is important to note that SSC is also, in many cases, tied to the accomplishment of national security, trade, investments and even international recognition goals of these new donors (Sanahuja 2010; Ayllón Pino & Costa Leite 2010; Sidiropoulos 2012).

Commentators and observers in the field of international political economy agree that emerging economies have projected themselves beyond their national boarders as a way to obtain new sources of natural resources that support and sustain their own domestic economic growth. When taking a closer look at the strategies developed by the BRICS in Africa, a recent report by the United Nations

Economic Commission for Africa recognises that 'key features of BRICS aid to Africa (particularly China, and to some extend India and Brazil) is [the] use of official flows to promote trade and investment' (UNECA 2013: 17). The report also highlights that even if there are differences among the actions developed by the BRICS in Africa (China's engagement in Africa being primarily state-driven with a strong focus on loans vs Brazil's emphasis on technical cooperation), trade and investments by BRICS in the continent seem to be locking Africa into a specialisation in primary commodities.

In this sense, Africa as well as other resource-abundant regions in the Global South, is becoming part of a new extractivist dynamic pushed by the contemporary international success of the 'commodification of nature' now also fostered by the actions developed by emerging economies. To a certain degree, a correlation between IC initiatives at a South-South level and a new extractivist orientation promoted by emerging economies countries could then be drawn.

The concept of extractivism refers to those economic activities based on the extraction of large volumes of natural resources with the goal of being exported without any or hardly any local processing (Gudynas 2010; Acosta 2011). Usually associated with mining or hydrocarbon-related activities, extractivism also includes other sectors such as the agrarian, forestry and fishery. Twenty-first century extractivist practices have called the attention of scholars giving origin to the concept of 'neo-extractivism'. Although thought for a South American context, this term introduced by Gudynas (2009), can be transposed to Africa shedding light over recent developments revolving around natural resource-based economic activities in that continent.

Within SSA, neo-extractivism in the agricultural sector can be summarised in the hazard introduced by large-scale agriculture, partly responsible for deepening the problematic phenomenon of *land grabbing*. This concept has been widely used in the fields of agrarian studies and political ecology and it generally characterises a process of appropriation of large sections of land by foreign capital (Taylor & Bending 2009; Merlet 2010; Sauer & Pereira Leite 2012) or by local landlords as pointed out by Borras et al (2012) for the case of Latin America. According to a World Bank report (2011), the 2007-2008 period of high and volatile prices on food products led to a new wave land demand with approximately 56 million hectares destined to new large-scale farmland deals announced between 2008 and 2009. More than 70 per cent of the world's demand for land corresponds to Africa, particularly in countries like Ethiopia, Mozambique and Sudan where millions of hectares of land have been transferred to investors.

Several global developments such as the international food crisis, CC and the growing demand for biofuels are among the main triggers of land grabbing by transnational companies. Even if in SSA countries the biofuel sector is at an incipient degree of development, there is a close nexus between land and energy investments. Large international land acquisition deals with the objective of growing energy crops have become one of the main forms of investments in the agrarian sector. This sudden race by international investors for closing up land deals in many countries in SSA, is naturally leading to numerous land conflicts (UNECA 2012) since land is a central asset in supporting the livelihoods of local populations but also one of the main sources for national development.

Contemporary Liquid Biofuel Developments in Sub-Saharan Africa

From the SSA countries' governmental perspective, there are three main drivers for promoting biofuels in the region: (1) The possibility of energy self-sufficiency in countries very much dependent on oil imports with potentials for economic, social and environmental benefits; (2) The chance of enhancing national savings in foreign currency through biofuel exports; (3) The opportunity for job creation and rural development (Von Maltitz et al 2008; Amigun et al 2011).

Most countries in SSA are characterised by poor institutional capacities. In many cases, the absence of well-designed biofuel legal and sustainability frameworks is the main impediment for achieving the objectives around which biofuels have been promoted in SSA (Duvenage et al 2012). Recent studies (Jumbe et al 2013; UNECA 2008) highlight an almost generalised absence of governmental policy instruments oriented to support and promote the development of biofuels in several countries within the region. Additionally, weak ties between domestic elites and the State make societies in these countries more exposed to the business interests of transnational corporation (Duvenage et al 2012).

Consequently, this shortcoming has been identified as an opportunity by external actors. Foreign investors, including the European Union, USA and Japan, are driving most SSA biofuel projects (Mshandete 2011; Amigun et al 2011). Despite the dominant presence of government and private sector biofuel actors from the Global North in SSA, it is worth highlighting the growing relevance of emerging actors as resource-seeking investors. As we will later see, SSC stands as a very functional 'letter of introduction' for achieving that goal.

A recent study that reviews liquid biofuel strategies in 13 SSA countries reveals that jatropha, sugarcane, canola and sweet sorghum are among the most common feedstock used in SSA for the production of either biodiesel or bioethanol (PISCES 2011). *Jatropha Curcas* was introduced as one of the most promising feedstock for biodiesel production in countries like Senegal, Mali, Burkina Faso, Benin, Mozambique and Zimbabwe which had already developed tasks forces for the promotion of this crop. However, it is important to highlight that while some countries have propelled the production of jatropha for biodiesel, South Africa has placed this crop on a list of invasive species (Amigun et al 2011).

Vermuelen et al. (2011) point out that many governments have created investment promotion agencies oriented towards targeting foreign direct investments and facilitating land access to transnational companies in the agricultural sector or producing biofuel feedstock. This is taking place in a complex context of competition between protecting customary land rights and ensuring land for large-cale agricultural investors. The 'investment imperative' is the prioritised position by several governments in Africa, being high (and not marginal) value land the most commonly subjected to international acquisition deals.

Additionally, there is a consensus among experts that biofuel developments n Africa should have a strong rural development component rather than a large-cale corporation/commercial focus (Giovanetti et al 2012; Jumbe et al 2013). However, the vast majority of the biofuel developments happen to be dominated by agro-industrial projects characterised by the acquisition of large portions of and for growing, processing and distributing bioenergy feedstock with actually very little local community involvement.

In the particular case of Senegal, the rush for biofuel production brought along a series of conflicts around the occupation of land that translated into strong confrontations between international investors and local communities. According to several governmental officials, foreign investors would directly enter into negotiations with the rural committees and purchase huge portions of land (between 10 and 20 thousand hectares) to develop energy crop projects. Initially these investors settled down in the northern part of Senegal causing a series of conflicts, which led many investors to move to the Luga region where new conflict with the rural communities arose.

Biofuel projects are already underway in several SSA countries, such as Angola, Benin, Ghana, Mozambique, Kenya, Mali, Malawi, Nigeria, Mozambique, Senegal, South Africa, Tanzania, Zambia and Zimbabwe even when this development is not necessarily enhancing the livelihoods of local populations and fail to be implemented in an environmentally sound fashion (Diaz-Chavez et al 2010; Richardson 2011; Hunsberger 2011; Duvenage et al. 2012).

In parallel at global level, the promise of liquid biofuels as key a driver for rural development, energy security and CC combating by introducing a source of energy able to reduce emissions, has begun to fade. Recently, a set of ever-evolving debates concerning global and regional issues such as food security, land use change, deforestation, land concentration and other relevant environmental and social impacts are raising doubts on the need to support biofuel policies at the governmental level in some SSA like Senegal.

Why the Sugarcane Tastes so Good: Analysing the Presence of Brazilian Biofuel-related Actors in Senegal

As mentioned before, this article seeks to explore the nexus between IC actions developed by Brazil as an emerging country and sustainable biofuel production in SSA countries by focusing on Senegal as a case study. Despite the strong technical component found in SSC, South-South initiatives fostered by Brazil in SSA countries for the development of biofuels are not strictly restricted to knowledge transfer and research but also include a trade and investment component. For thi reason, we reckon that assessing the political, technical and economic dimension of IC is necessary for understanding Brazil's role in Senegal. By focusing on al these dimensions, we will be able to provide a full image of the internationa involvement of Brazilian actors in the development of biofuels in Senegal.

This section's analysis draws on interviews with key informants in Brazi and Senegal, including representatives of the Brazilian Ministry of Internationa Relations (*Itamaraty*), the Brazilian Enterprise for Agricultural Research (*EMBRAPA Agroenergia*), the Senegal's Ministry of Energy, the National Agency of Renewable Energies, the Senegalese Institute for Agricultural Research (ISRA and the Agency for the Promotion of Investments of Senegal (APIX).

As one of the major challenges of this work was how to obtain officia statistics, press reports on investments and production released by the Braziliar national media were used when necessary to support the arguments. Finally, the strengthening of Brazilian ties with Senegal is the result of actions developed no only by governmental actors and organs, but also by decentralised semi-public anc private actors that have acted under distinct logics and following different objective: and priorities. With the objective of reflecting Brazil's new IC actions for biofue development in SSA countries, we have focused in the following section on a se of public, semi-public and private actors that have played a key role in promotinε biofuel-related cooperation actions, research, policies and investments.

The Ethanol Diplomacy: A Look at the Political Dimension of Brazilian SSC with Senegal

Bilateral political relations between the governments of Brazil and Senegal were the outcome of Brazil's more general ambition to projecting itself as global promoter o developmental cooperation, following its international commitments[2] to suppor less developed countries in the Global South. By 1972, Brazil had signed a General Technical Cooperation Agreement with Senegal, which aimed to foster cooperation in agriculture.[3]

It is worth highlighting that the search of a political alliance between Brazil and Africa became significantly more intense and fluid after the arrival in 2003 of formei president Luiz Inácio 'Lula' da Silva. Since then SSC has crystalised into one of the central axis of Brazil's foreign action, with cooperation with Africa involving ovei 170 institutions among federal government organs (IPEA/ABC 2013) as well as other Brazilian private institutions (non-governmental organisations, foundations and corporations) that act on a wide array of areas, including education, health, urban and rural development, agriculture, environment, energy, among others.

In this new context of Brazil-Africa cooperation, despite Brazil's evident geographical and climatic advantages over Senegal, the assumption of shared agroclimatic conditions was an element retrieved in the political discourse for engaging in actions with this and other SSA countries. Agriculture is a key component in the economic development strategy of Senegal, therefore, the Brazilian government presented itself as a rich source of experience and knowledge that could be applied to Senegal contributing in this way with the African country in the achievement of Millennium Development Goals.

Along Lula's second mandate, and during the visit of Senegal's former president Abdoulaye Wade to Brazil in May 2007, a specific bilateral bioenergy agreement was signed between the two countries as a way for Brazil to foster the production of energy crops in that African country[4]. Senegal's high dependency on imported fossil fuels was the main driver that led that government to sign that agreement as Brazil's experience in the sector motivated the birth of a biofuel policy in Senegal during that same year. According to a representative of Senegal's Renewable Energy National Agency, 'soon after that trip to Brazil, the former president decided to encourage local development of biofuels (...). Later that year, a Brazilian delegation came to Senegal to determine the potential in terms of land, climate, rainfall regime and temperature for the development of energy crops. This visit conducted by the Brazilian Ministry of External Relations decided that Senegal gathered the necessary conditions for the development of biofuels'[5].

The goal in this African country is that biofuels are to be used as 'a partial or total substitute for fossil fuels' in the transport sector. Additionally, the government of Senegal highlighted that biofuels projects 'contributing to the reduction of greenhouse gas emissions, could be object of certification under the Clean Development Mechanism'[6].

The gesture to legally institutionalise the political cooperation with Senegal in the biofuel sector, in correlation with a wave of similar agreements through SSA, can be understood as a tool aimed to reinforce Brazil's 'Ethanol Diplomacy'. The Ministry of External Relations, took the lead in translating the political agreements negotiated by Lula's 'Ethanol Diplomacy' into concrete technical cooperation actions by launching in 2009 the 'Structured Support Programme to other Developing Countries in the area of Renewable Energies' (Pró-Renova). This programme, domiciled in the Division for New and Renewable Energy of Itamaraty, opened up the way to implementing technical-scientific cooperation actions related to the development of biofuel production feasibility studies in 24 different African countries (Government of Brazil 2011).

A Biofuels Model for Africa: The Technical-Scientific Dimension of Brazilian SSC

Around 2007, the international donor community spread the promise that jatropha curcas was the ideal energy crop for many countries in SSA. Following the assumption that this miracle crop does not demand major quantities of water to grow, being therefore able to grow in semi-arid areas where precipitations are scarce, the Senegalese government selected jatropha as the source for producing biodiesel for the transport sector.

The plan launched in 2008 promoted the use of jatropha on a small scale involving the rural communities. The goal was for each community to develop 1.000 hectares of this crop. ISRA (Senegalese Institute for Agricultural Research) was the technical partner in charge of developing and distributing the plant to all communities free of costs. Following this plan, and as there are about 360 organised rural communities in Senegal, the government expected to develop 360,000 hectares of jatropha in a period of 4 years (2008-2012). However as previous experience with this plant was inexistent, the plan failed. According to one of the interviewees, 'despite everything that was once said about this miraculous plant, it proved to be a failure from the moment that plantations started evolving'.

This failure is not an isolated fact. Other countries in West Africa that promoted jatropha such as Mali, Benin and Burkina Faso shared similar problem where the experience showed that even if the plant is able to grow in marginal or degraded land, the yields are too low. Consequently, in Senegal, the installed jatropha crushing plants that settled down in the *Gossace* region did not have enough grains in order to produce the necessary oil for producing biodiesel; hence production had to stop.

This example illustrates what several scholars have observed in the field: in order to tackle the potential of biofuels in SSA countries and ensure that international interest do not contradict national objectives associated with their introduction, there is an urgent need to build capacities, technical skills in the agrarian and industrial phases and to develop biofuels along clear legislative and regulatory frameworks (Jumbe et al 2009; Amigun et al. 2011; Jumbe et al. 2013).

These needs were quickly identified as an opportunity by the Brazilian government. In the frame of its bilateral cooperation arrangement with the Republic of Senegal, a set of Brazilian governmental and non-governmental technical-scientific organisations became intensively involved in trainings, capacity building and biofuels development associated research in Senegal as well as in several other SSA countries.

The governmental body in Brazil responsible for articulating the position of their ministries and other national institutions and providing the necessary technical support to implement its international cooperation agreements is the Brazilian Cooperation Agency (ABC). According to the last official report on Brazilian-provided technical cooperation (COBRADI 2013), Africa was the second largest recipient of Brazilian technical cooperation (39.4 %) after Latin America and the Caribbean (53.3 %). Among the most demanded areas of technical cooperation by Africa, agriculture takes the largest percentage of the developed actions by ABC in the continent (ABC 2011).

When technical cooperation involves agriculture-related capacity building and knowledge transfer actions, ABC is supported by the technical-scientific know-how of the Brazilian Enterprise for Agricultural Research (EMBRAPA). This state-owned company created in 1973 as part of the Ministry of Agriculture, was responsible for the productive revolution that transformed the Brazilian *Cerrado* biome into the new core of soybean and sugarcane production. This process is associated with the introduction of genetic innovations for obtaining high yields under tropical climatic conditions (EMBRAPA 2013). Since then, EMBRAPA has been able to become one of the main globally recognised public research centres specialised in tropical agriculture and bioenergy. This advantage has eased the internationalisation strategy of the company, leading to its opening representation offices in Panama and Venezuela (Latin America) and Ghana (Africa). From the Accra office in Ghana, EMBRAPA coordinates over 51 agriculture-related projects in Africa (IPEA/BM 2011).

Taking advantage of multilateral bioenergy cooperation platforms, such as Global Bioenergy Partnership (GBEP), the Brazilian Ministry of External Relations co-organised with EMBRAPA and the Food and Agriculture Organization (FAO) the 'Bioenergy Week' in Brasilia in early 2013. Through capacity building trainings, examples of success in Brazil's bioenergy sector were diffused to a large delegation of SSA representatives from ECOWAS (Ghana, Senegal, Gambia, Guinea-Bissau, Mali, Niger, Togo, Ivory Coast, Cape Verde) and Mozambique.

Another relevant non-governmental actor is *Fundação Getúlio Vargas*. Through its unit Projetos (*FGV Projetos*) it played a key role in providing technical-scientific expertise in the frame of the *Pró-Ronova's* actions developed by the Brazilian government in SSA countries. FGV Projetos is the unit for technical consultancy of *Fundação Getúlio Vargas*, a prestigious Brazilian private higher education institution/ think tank designed to promote Brazil's economic and social development.[7] The unit has been involved in several bilateral and trilateral cooperation initiatives promoted by Brazil, developing technical feasibility studies for the production of biofuels in SSA.

FGV Projetos is currently carrying out the Project 'Biofuels Production- FGV Foundation' a study for the agricultural potential of six SSA countries located in the tropical belt.[8] Senegal was one of the first SSA countries to integrate the group of beneficiaries of Brazilian technical cooperation in the biofuel sector. As the government of Senegal was looking to develop a biofuel policy and regulatory framework, a trilateral cooperation arrangement between Brazil, the United States and Senegal allowed the technical intervention of *FGV Projetos*. The Brazilian research centre was responsible for carrying out an economic-financial-technical biofuel feasibility study in 2010 with the technical support of *ISRA* at the request of *Itamaraty* who financed the project. The study 'reinforces the feasibility of introducing biofuels into the Senegal energy matrix and the capacity to attract private investments (…) Africa, in turn, is emerging as a promising large-scale biofuel producer, considering the existence of large areas of arable land, the tropical climate and available man power' (FGV *Projetos* 2010:4). According to the cited document, an agricultural zoning technical study was developed for determining the feedstock with the highest potential for biofuels: sugarcane, soy, cotton and sunflower were indicated as the recommended feedstock for producing bioethanol or biodiesel. Consequently, three projects were recommended in the country: (1) A 3,000-hectare project for sugarcane bioethanol in the Tambacounda department; (2) A 2,500-hectare soy and sunflower biodiesel project in the Zighunchor region; and (3) A 3,600-hectare project for the Kaolack region (FGV Projetos 2010). Government representatives interviewed on these projects highlighted that in the frame of this initiative, rural communities had agreed with the national government to provide 3,000 hectares for a sugarcane pilot project, as the echo of molasses sugarcane bioethanol has become much stronger after the jatropha experience.

By December 2010, a Biofuel Law[9] was enacted by the president of the Republic. However the law requires a decree of application in order to set a price structure, a mandatory blend, etc. Nonetheless, after a change of president in the country, the decree has been awaiting parliamentary approval since 2012. As this decree would also serve as a guarantee for international investors to rely on basic game rules in the sector in Senegal, interviewees highlighted how the Brazilian government has been very active at lobbying through their local embassy and holding interviews with Senegal's Prime Minister for the decree's final approval.

The Economic Dimension of Brazilian SSC

Investments in infrastructural sector has also been identified as yet another precondition for the development of a biofuel industry in SSA as a way for these countries to further grasp the benefits of the growing international biofuel markets. Even if in terms of economic cooperation Brazil rarely provides concessional loans – emphasising instead scientific-technical cooperation and technological transfers (UNECA 2013), the Brazilian state does subsidise both its state and privately-owned companies acting in SSA.

The Brazilian National Bank for Economic and Social Development (BNDES) was created in 1952, thus becoming the main financing institution of the federal government. It has been playing an active role in supporting external trade and the internationalisation of Brazilian companies.[10] The strategy to internationally foster Brazilian 'national champions' started early in the 2000 decade with the fusion with and/or acquisition of large companies in sectors such as energy, mining and food in several Latin American countries.[11] Owing to the successful experience in Latin America, in 2008 BNDES created its International Department responsible for the international actions carried out by the bank. This initiative was complemented by he inauguration of regional offices for Latin America (Montevideo) and in 2009 n Europe (London). More recently, in late 2013, a new representation office for Africa was opened in Johannesburg (South Africa).

In the search of expanding Brazil's economic presence beyond South Africa and the Portuguese-speaking countries of the continent, a number of official visits to SSA countries have been organised since 2009 by the Ministry of Development, Industry and External Trade (MDIC) with special focus on Western Africa. One of the countries the former minister visited personally was Senegal; and he did this to promote bilateral trade and Brazilian investments in this country.

In relation to the latter, the decision of President Dilma Rousseff to cancel or renegotiate up to US$ 900 million of the debts of several African countries to Brazil in 2013 is also part of that plan of expanding the frontiers of Brazilian investments to other emerging African markets. Until recently, Brazilian banks, such as BNDES, were not able to finance investment and trade to those countries due to the existence of debts with the Brazilian government. Therefore, during Lula´s government, Brazilian economic cooperation through BNDES in SSA was restricted to countries such as Angola, Mozambique and South Africa where BNDES has been financing the export of Brazilian capital goods since 2007 (Government of Brazil 2010; Motta Veiga 2013). With this recent decision by the current Brazilian President, Dilma Rousseff, 12 new SSA countries including Senegal will be benefitting from new investments. In this sense, the Brazilian Ministry of Agrarian Development and the government of Senegal have agreed on a credit line for the purchase of Brazilian agrarian machinery and equipment.

This expansion of Brazilian capital to Senegal and other SSA countries over the last few years can be understood as the latest phase of internationalisation of Brazilian companies promoted by BNDES. In some countries such as Angola and Mozambique, the bank also participates in promoting investments for actors directly involved in the biofuels productive chain. Owing to the success of this experience, *Itamaraty* and BNDES formalised their collaboration by signing in 2011 a cooperation agreement for promoting biofuels in other developing countries with strong focus in SSA.[12]

These elements provide a hint of the Brazilian government's interest in developing an ethanol market in Western Africa. However, in the Senegalese context in which the absence of an approved decree is a barrier to implementing the biofuel national law, Brazilian official investments in the biofuel sector are automatically placed on hold. According to an official of Senegal's Ministry of Energy, financing for the development of sugarcane-based ethanol are ready to be implemented through the trilateral cooperation agenda between Senegal, Brazil and the USA. A Memorandum of Understanding between these countries is already signed for developing a project of 20 thousand hectares to produce about 60,000 million litres of ethanol, but the investments are still awaiting the biofuel normative framework approval of Senegal.

Meanwhile, the developments observed in the sector in Senegal are hardly part of a comprehensive public policy. According to some of the interviewees, the main goal of companies investing in the sector is not necessarily to supply the local market but rather to export oil and biofuels to external markets. In this wise, even when the law establishes that external biofuel producers in the country must leave at least 50 per cent of their production for consumption in the local market, as the law is not yet applicable, the external privately-owned companies already operating in the country tend to invest either in the agricultural phase exporting the produced raw materials or invest in the industrial phase with the objective of exporting the entire production overseas.

Concluding Remarks: Limits and Shortcomings of the Brazilian Model of Liquid Biofuel Expansion through International Cooperation

Through the case study of biofuels developments in Senegal, we have been able to explore and depict the behaviour of Brazil as an emerging donor country in SSA. The analysis of the three dimensions presents in Brazil's IC strategy, has allowed us to see that Brazil has the political ambition, technical-scientific know-how and the necessary economic capacity for developing new biofuel agro-industrial projects in Senegal as well as in other SSA countries. It has become the first South American country to play an active role in IC outside its own region. The number of technical assistance initiatives and the volume of resources provided by the State, in association with other non-state agents, also reveals an original element in the new path the country is forging as an international emerging actor.

Nonetheless, we observed that behind the mission of South-South solidarity, there is an evident political goal in Brazil's international biofuel cooperation initiatives: the transformation of ethanol into a global commodity. The larger the number of countries involved in the production of ethanol, the more chances Brazil has in succeeding in the achievement of such geopolitical objective. Therefore, the analysed developments regarding the Brazilian presence in Senegal reveal a complete correspondence with the announced objective of Brazilian *Plano Nacional de Agroenergia* to lead the creation of an international ethanol market. In that direction the unfolded 'Ethanol Diplomacy' during former president's Lula mandates, oriented towards facilitating the transfer of technical and scientific know how, has been a key tool for achieving that foreign policy goal.

The modernisation and opening of the Brazilian economy, as well as the analysed SSC initiatives, has propelled an international geographic expansion of the nationally consolidated biotechnological revolution in the agrarian sector. When looking at other SSA countries such as Angola and Mozambique, the size and nature of the transnationalised investments seem to indicate that the process led by the international expansion of Brazilian companies in the biofuel sector, could lead to a reproduction of a model of large-scale agribusiness in SSA countries.

Several of the key actors interviewed in Senegal highlighted the need to anticipate the risks associated with the diffusion of a large-scale agro-industrial productive model associated with sugarcane. Although successful and efficient for countries with a strong agroindustry like Brazil, the introduction of this model in countries where family agriculture is by far the dominant reality could have disastrous effects. Therefore, the gathered evidence shows that biofuels as a tool for dealing with CC seem to be introducing deep asymmetries and inequalities between more powerful and less privileged actors and sector also at a South-South level.

In countries such as Senegal, characterised by weak state structures and underdeveloped regulatory mechanisms, the main question arising is whether an emerging actor like Brazil will be able to keep the balance between using and adapting its expertise, knowledge, public and private funds for the provision of a public good or for supporting particular sectoral interests in the production of biofuels. Against the backdrop of tis political economy, the socio-environmental sustainability of this development has recently become a global topic of debate, especially after the ratification of the EU Renewable Energy Directive in 2009.

We highlighted how the promotion of liquid biofuels as a source of energy in Africa is showing severe impacts such as the degradation of fragile ecosystem, the displacement of rural workers and populations and diversion of the agrarian production. These risks are responsible for a deep and generalised discredit in biofuel policies at a global level, which results in two different but interrelated outcomes. On the one hand, in the particular case of Senegal, this factor appears as one of the main causes behind the current paralysis in the implementation of the national biofuel policy. On the other, that same global discontent with the evolution of liquid biofuel developments also challenges the 'Ethanol Diplomacy' model that the Brazilian government has been offering to its SSA partners.

Notes

1. The concept was first defined in the Accra Agenda for Action on Aid Effectiveness in 2008 as a sort of cooperation that involves developing countries in the search of improving development conditions in Third World countries on the basis of principles such as non-interference in internal matters, equality among partners, respect for sovereignty, cultural diversity, identity and local content (AAA 2008, section 19e). The Accra Declaration was oriented towards deepening the Paris agenda which emerged in 2005 with the Paris Declaration on Aid Effectiveness. This declaration proposed joint efforts by the international community in order to achieve the targets set in the Millennium Declaration (2000).

2. We refer to the 'Buenos Aires Action Plan' adopted during the United Nations Conference on Technical Cooperation among Developing Countries in 1978.

3. 'Basic Technical Cooperation Agreement between the government of the Federal Republic of Brazil and the government of the Republic of Senegal' available in: http://dai-mre. serpro.gov.br/atos-internacionais/bilaterais/1972/b_115/at_download/arquivo (accessed May 2014)

4. 'Ajuste Complementar ao Acordo de Cooperação Técnica entre Brasil e Senegal para a implementação do projeto Formação de Recursos Humanos e Transferência de Tecnologia para Apoio ao Programa Nacional de Biocombustíveis no Senegal'
5. Statement retrieved during fieldwork in Senegal (June 2014).
6. Articles 5 and 21. *Loi d'Orientation de la Filière de Biocarburants*. Law N° 2010-22. Government of Senegal, 15th December 2010.
7. For more information visit: fgvprojetos.fgv.br
8. Guinea, Guinea-Bissau, Liberia, Mozambique, Senegal and Zambia.
9. *Loi d'Orientation de la Filière de Biocarburants*. Law N° 2010-22. December 15th 2010.
10. For further information on BNDES international insertion strategy see the bank's *Exportação e inserção internacional* section in its official website: www.bndes.gov.ar (accessed March 2014)
11. For further analysis and more complete discussion on the role of BNDES in the internationalization of Brazilian companies in Latin America, see Perrotta et al. (2011).
12. '*Celebração de Acordo de Cooperação entre o Itamaraty e o BNDES para Promoção de Biocombustíveis em Países em Desenvolvimento*' Brasília 17 de fevereiro de 2011 in Brazilian Ministry of External Relations'official website: www.itamaraty.gov.br (accessed April 2014).

References

ABC, 2011,)*Brazilian Technical Cooperation*. Agencia Brasileira de Cooperação Internacional. Ministerio das Relações Exteriores. Brasília.

Acharya, A., 2014, *The End of the American World Order,* Cambridge: Polity Press.

Acosta, A., 2011,'Extractivismo y neo-extractivismo: dos caras de la misma maldición' en Miriam Lang & Dunia Mokrani (eds) *Mas allá del desarrollo. Grupo Permanente de Trabajo sobre Alternativas al Desarrollo.* Noviembre. Quito.

Amigun, B. & J. Musango, 2011, 'An Analysis of Feedstock and Location for Biodiesel Production in Southern Africa', *International Journal of Sustainable Energy,* Vol. 30, No. S1,Taylor and Francias.

Arndt, T., 2011, 'The case of Mozambique', Copenhagen: Department of Economics, Development Economics Research Group, University of Copenhagen.

Ayllon Pino, B., & I. Costa Leite, 2010, 'La cooperación Sur-Sur de Brasil. Proyección solidaria y política exterior. En B. Ayllón Pino & J. Surasky (coords.) *La cooperación Sur-Sur en Latinoamérica.* Utopía y Realidad. Madrid: Catarata,

Badie, B., 2013, *Diplomacia del contubernio. Los desvíos oligárquicos del sistema internacional.* Universidad Nacional Tres de Febrero.1º Edición, Buenos Aires.

Borras, S.; J. Franco, S. Gómez, C. Kay & M. Spoor, 2012, 'Land grabbing in Latin America and the Caribbean'. Journal of Peasant Studies Vol. 39, No 3-4: July-October.

Cabral, L. & A. Shankland, 2013, *Narratives of Brazil- Africa Cooperation for Agricultural Development: New Paradigmes?* Working Paper N° 51. Future Agricultures.

COBRADI, 2013, *Cooperação Brasileira para o Desenvolvimento Internacional: 2010.* Instituto de Pesquisa Econômica Aplicada/ Agência Brasileira de Cooperação. Brasília.

Daily, G., 1997, 'Introduction: What are ecosystem services?' in Daily, G., ed., *Nature's Services: Societal Dependence on Natural Ecosystems*, Washington, DC. :Island Press.

Delgado Ramos, G., 2013, *Biocombustibles en México. Cambio Climático, Medio Ambiente y Energía*. CEIICH/ PINCC, UNAM. México

Diaz-Chavez, R.; S. Mutimba, H. Watson, S. Rodriguez Sanchez & M. Nguer, 2010, *Mapping Food and Bioenergy in Africa. A Report prepared on behalf of FARA*. Forum for the Agriculture Research in Africa. Accra, Ghana.

DUVENAGE, I.; R. Tupin & L. Stringer, 2012, 'Towards Implementation and Achievement of Sustainable Development in Africa'. *Environment, Development and Sustainability* N° 14.

Embrapa, 2013, *Quarenta Anos de Construçoes*. Empresa Brasileira de Pesquisa Agropecuaria. Ministerio de Agricultura, Pecuária e Abastecimento.

Embrapa Agroenergía, 2013, 'Semana da Bioenergia'. Agroenergético: Informativo da EMBRAPA Agroenergia N°40, Abril. Brasília.

ESPA, 2013, *Sugar Rush*. London: Ecosystem Services for Poverty Alleviation, July.

Falkner, R., 2014, 'Global Environmental Politics and Energy: Mapping the Research Agenda' *Energy Research and Social Sciences* Nro. 1. March.

Ferreira De Lima, J. (2012) 'A ABC e a EMBRAPA na Africa. Parceria na Cooepracao Técnica: O Caso de Mozambique'. Instituto de Ciencia Política e Relacoes Internacionais. Universidade de Brasilia. Brasilia.

FGV *Projetos*, 2010, *Biofuel Production in the Republic of Senegal. Stage 1: Feasibility Study. Summary of the Final Report of Stage 1*. Fundação Getúlio Vargas. Novermber.

FGV *Projetos*, 2013, *Brazilian Sustainable Methodology for Bioenergy Production in Africa*. Fundação Getúlio Vargas. Presentation by FGV Projetos officer during Biofuels Workshop in Maputo, Mozambique. March.

FoE, 2010, *Jatropha: Money doesn't grow on trees. Ten reasons why jatropha is neither a profitable nor sustainable investment*. Friends of the Earth International. Issue 120. December.

Franco, J., L. Levidow, D. Fig, L. Goldfarb, M. Hönicke & M. Mendonça, 2011. 'Assumptioons in the European Union biofuels policy: frictions with experiences in Germany, Brazil and Mozambique', in Borras, Saturnino et al., eds, *The Politics of Biofuels, Land and Agrarian Change: Critical Agrarian Studies*, New York: Routledge.

Fulquet, G., 2015, '¿La maldición de los recursos naturales? Conocimiento experto, política e intereses sectoriales en el desarrollo de biocombustibles en Sudamérica'. Brazilian Journal of International Relations, Vol. 4, Edição N°1

Gasparatos, A., P. 'Stromberg & K. Takeuchi, 2013, 'Sustainability Impacts of First-Generation Biofuels' *Animal Frontiers*, Vol. 3, N°2, April.

German, L. G. Schonevled & E. Mwangi, 2011, *Contemporary Process of Large-Scale Land Acquisition by Investors. Case Studies from sub-Saharan Africa*. Center for International Forestry Research CIFOR. Occasional Paper 68.

GIOVANETTI, G. & E. Ticci, 2012, 'Biofuel Developments and Large-Scale Land Deals in Sub-Saharan Africa'. Economics of Global Interactions Conference. Universita degli Studi di Bari.

Government of Brazil, 2010, 'Balanço de Politica Externa 2003-2010. Relações com a África. Comércio e investimentos'. Ministério das Relações Exteriores, Itamarary. Dezembro, Brasília.

Government of Brazil, 2011, 'Temas Multilaterais: Energia. Energias renováveis, incluindo biocombustíveis'. Ministério das Relações Exteriores, Itamarary. Fevreiro, Brasília.

Government of Mozambique, 2009, *Política e Estratégia de Biocombustiveis .Resolução N°2 22/2009*. Boletim da República. Publicacao Oficial da República de Mocambique. 3° Suplemento. 21 de Maio.

Gudynas, E., 2009, 'Diez tesis urgentes sobre el nuevo extractivismo. Contextos y demandas bajo el progresismo sudamericano actual'. Centro Andino de Acción Popular y Centro Latinoamericano de Ecología Social. Disponible en: http://biblioteca.hegoa.ehu.es/system/ ebooks/17745/original/Gudynas_Nuevo_Extractivismo_10_Tesis.pdf

Gudynas, Eduardo, 2010, 'Agropecuaria y nuevo extractivismo bajo los gobiernos progresistas de América del Sur.' Revista Territorios 5: 37-54. Instituto de Estudios Agrarios y Rurales – CONGCOOP, Guatemala.

Hancock, K. & V. Vivoda, 2014, 'International Political Economy: A Field Bron of the OPEC Crisis Returns To Its Energy Roots'. *Energy Research and Social Sciences* Nro. 1. April.

Hochstetler, K., 2012, 'Brazil as an Emerging Environmental Donor'. *Policy Brief* No. 21. The Center for International Governance Innovation. February.

Humphrey, J. & D. Messner, 2006, *Unstable Multipolarity? China's and India's Challenges for Global Governance*, Berlin: German Development Institute.

Hunsberger, C., 2011, 'The Politics of Jatropha-based biofuels in Kenya: Convergence and Divergence among NGOs, Donors, Government Officials and Farmers', in Borras, S.; P. McMichale & I. Scoones, eds, *The Politics of Biofuels, Land and Agrarian Change: Critical Agrarian Studies,* New York: Routledge.

Hurrell, A., 2007,. *On Global Order: Power, Values and the Constitution of International Society.* Oxford: Oxford University Press.

Lechini, G., 2011, *Argentina and South Africa Facing the Challenges of the XXI Century. Brazil as the Mirror Image.* Editorial de la Universidad Nacional de Rosário. 1a Edición. Rosario

IPEA/BM, 2011, *Ponte sobre o Atlântico. Brasil e África Subsaariana: parceria Sul-Sul para o crescimento.* Instituto de Pesquisa Econômica Aplicada/ Banco Mundial. Dezembro 14. Brasília/ Washington DC.

Jumbe, C. & M. Mkondiwa, 2012, 'Comparative Analysis of Biofuels Policy Development in Sub-Saharan Africa: The Place of Private and Public Sectors'. *Renewable Energy: An International Journal.* Vol.50

Lafer, C., 2002, *La identidad internacional de Brasil. Fondo de Cultura Económica (FCE)*, Buenos Aires.

Lengyel, M. & B. Malacalza, 2011, 'What Do We Talk When We Talk About South-South Cooperation? IPSA-ECPR Joint Conference What happened to North-South? Sao Paulo. Februaru.

MAPA, 2006, *Plano Nacional de Agroenergía 2006-2011*. 2da Edição Revisada. Ministerio de Agricultura, Pecuaria e Abastecimento.

MRE, 2011, *Programa Estruturado de Apoio do Brasil aos demais Países em Desenvolvimento na Área de Energias Renováveis PRO-RENOVA: Apresentação.* Divisão de Fontes de Energias Novas e Renováveis. Ministério das Relações Exteriores, Brasília.

Mshandete, A., 2011, Biofuels in Tanzania: Status, opportunities and Challenges', *Journal of Applied Bioscience* Vol. 40.

Merlet, M., 2010, *Les appropiations des terres a grande échelle: analyse du phénomene et propositions d'orientations.* Comité Foncier et Développement. June, Paris.

Milani, C., 2014, *Brazil's South-South Co-operation Strategies: From Foreign Policy to Public Policy*. Occasional Paper No. 179. Global Powers and Africa Programme. South African Institute of International Affairs. March.

Motta Veiga, P., 2013, 'A África na agenda econômica do Brasil: comércio, investimentos e cooperação'. Revista Brasileira de Comercio Exterior N° 116. Julho-Setempro. FUNCEX.

Nederveen Pieterse, J., 2008, 'Globalization the Next Round: Sociological Perspectives', *Futures Journal* No. 40.

Nolte, D., 2008, 'Ideas, Interests, Resources and Strategies of Regional Powers'. Paper presented at the 1st Regional Powers Network (RPN) Conference on 'Ideas, Interests, Resources and Strategies of Regional Powers. Analytical Concepts in Comparative Perspective'. German Institute of Global and Area Studies, September. Hamburg, Germany.

OXFAM, 2012, *The Hunger Grains. The Fight Is on. Time to scrap EU Biofuel mandates*. OXFAM Briefing Paper 161. September.

Pelfini, A.; G. Fulquet & A. Beling, 2012, *La Energía de los Emergentes. Innovación y cooperación para la promoción de energías renovables en el Sur Global*. Teseo/FLACSO, Buenos Aires.

Perez De Aramiño, Karlos, 2000, Diccionario de Accion Humanitaria y Cooperacion al Desarrollo. Universidad del País Vasco. Editorial Icaria. (accedido online en http://www.dicc.hegoa.ehu.es/ 24 de Mayo 2013)

Perrotta, D.; G. Fulquet & E. Inchauspe, 2011, *Luces y sombras de la internacionalización de empresas brasileñas en Sudamérica: ¿Integración o Interacción?* Documentos Nueva Sociedad, Friedrich Ebert Stiftung. Buenos Aires. Enero.

Pisces, 2011, *Liquid Biofuel Strategies and Policies in selected African Countries. A review of some of the challenges, activities and policy options for liquid biofuels*. Policy Innovation Systems for Clean Energy Security/ University of Edinburgh.

Richardson, B., 2011, 'Big Sugar in Southern Africa: Rural Development and the Perverted Potential of Sugar/Ethanol Exports,' in Borras.S; P. McMichale & I. Scoones, eds, *The Politics of Biofuels, Land and Agrarian Change: Critical Agrarian Studies*, New York: Routledge.

Sanahuja J., 2010, 'La construcción de una región: Suramérica y el regionalismo posliberal', in Cienfuegos, M. Y Sanahuja J., coordinators, *Una región en construcción. UNASUR y la integración en América del Sur*, Barcelona, Fundació CIDOB.

Saraiva, J., 2012, *África parceira do Brasil atlântico: Relações Internacionais do Brasil e da África no inicio do século XXI*. Coleção Relações Internacionais, Serie Parcerias Estratégicas com o Brasil. Fino Traço, Belo Horizonte.

Sauer, S. & S. Pereira Leite, 2012, 'Agrarian Structure, Foreign Investment in Land and Land Prices in Brazil'. *The Journal of Peasant Studies*, Vol. 39, issues 3-4.

Schlesinger, S., 2012, *Cooperação e Investimentos Internacionais do Brasil. A Internacionalização do etanol e do biodiesel*. Federação de Órgãos para Assistência Social e Educacional. FASE. Julho.

Schut, M.; M. Slingerland & A. Locke, 2010, 'Biofuel developments in Mozambique: Update and Analysis of Policy, Potencial and Reality', *Energy Policy*, Vol. 38. Elsevier.

Seabra, P., 2011, 'Brazil Upward Spiral: From Aspiring Player to Global Ambitions', in Tzifakis, N., ed., *International Politics in Times of Change*, Heidelberg: Springer.

Sidiropoulos, E., 2012, 'Rising Powers, South-South Co-operation and Africa', *Policy Briefing* No. 47. Global Powers and Africa Programme. South African Institute of International Affairs. March.

Smith, K., 2014, 'The BRICS alternative: Implications for Africa and the Global South' Paper presented at the FLACSO-Argentina/ CLACSO Joint Conference Rise and fall of international powers: an Assessment of the BRICS. ISEN. Argentinean Ministry o International Relations, Buenos Aires, 30 October.

Siebert, G., 2011, *Brazil in Africa: Ambitions and Achievements of an Emerging Regional Powe. in the Political and Economic Sector.* Centro de Estudos Africanos. Instituto Universitátic de Lisboa.

Taylor, M. & T. Bending, 2009, *Increasing Commercial Pressure on Land: Building a Coordinatec Response,* Rome: International Land Coalition.

UNCTAD, 2010, *Economic Development in Africa Report 2010. South-South Cooperation Africa and the New Forms of Development Partnership,* Geneva: United Nations Conference on Trade and Development.

UNDP, 1994, *The Buenos Aires Plan of Action,* Special Unit for TCDC, New York: United Nations Development Programme, November.

UNECA, 2008, *Biofuels: What strategies for Developing the Sector in West Africa?* United Nations Economic Commission for Africa. Bureau Sous Regional pour l'Afrique de l'Ouest. Nyamey, Niger.

UNECA, 2012, *Natural Resources & Conflict Management. The Case of Land,* Addis Ababa Ethiopia:

United Nations Economic Commission for Africa in collaboration with the Land Policy Initiative. UNECA, 2013, *Africa-BRICS Cooperation: Implications for Growth, Employmen. and Structural Transformation in Africa,* Addis Ababa, Ethiopia: United Nations Economic Commission for Africa.

UNU, 2012, *Biofuels in Africa. Impacts on Ecosystem Services, Biodiversity and Human Well-Being,* Yokohama: Institute of Advanced Studies, United Nations University.

Vermuelen, S. & L. Cotula, 2011, 'Over the Heads of Local People: Consultation, Consen• and Recompense in Large-Scale Deals for Biofuels in Africa', in Borras.S; P. McMichale & I. Scoones, eds, *The Politics of Biofuels, Land and Agrarian Change: Critical Agrarian Studies. New York: Routledge.

Von Maltitz, G & A. Brendt, 2008,. *Assesing the Biofuel Options for Southern Africa.* Science Real And Relevant: 2nd CSIR Biennial Conference, 17-18 November, Pretoria, South Africa.

World Bank, 2011, *Rising Global Interest in Farmland. Can it Yield Sustainable and Equitable Benefits?* The International Bank for Reconstruction and Development, Washington, D.C.: The World Bank.

6

Climate Change and the Urban Political Ecology of Water

Gian Carlo Delgado-Ramos

Introduction

Today, 52 per cent of the world's population live in urban areas. By 2050, this figure is expected to rise to between 64 and 69 per cent of the total world population (United Nations 2011). By this time, the size of urban areas is expected to have doubled or even tripled, depending on population and economic dynamics (Angel et al. 2011; IPCC 2014).

Cities consume between 67 per cent and 76 per cent of total world energy and are responsible for 71 per cent to 76 per cent of direct and indirect greenhouse gas (GHG) emissions (IPCC 2014). However, alone, the 380 developed region cities in the top 600 by GDP accounted for 50 per cent of global GDP in 2007 with more than 20 per cent of global GDP coming from 190 North American cities alone (McKinsey Global Institute 2013: 1). This positions such cities as practically the greatest consumers on the planet while the rest of urban settlements still play a minor role.

The foregoing is supported by the fact that although human urban settlements are growing at a rate of approximately 2 per cent per year, with outliers of 0.7 per cent for some developed countries and 3 per cent for some developing areas (United Nations 2011), this growth is not proportional to the amount of emissions that can be attributed to each case. Currently, similar urban settlements (by densities or the number of inhabitants per km^2) have very different GHG contributions – both historical and nominal (see Figure 1 for a comparative analysis of nominal emissions). Although, on the one hand, this divergence occurs partially in response to various factors such as land use, settlement form and extension, the length of time a settlement has existed, or the biophysical conditions of each case (e.g., latitude, proximity, and resource availability), on the other hand, the polarisation between cities and between inhabitants continues to be certainly significant not only in economic terms but also in the terms of energy and material consumption patterns.

Figure 6.1: Per capita GHG Emissions Versus Population Density for Selected Cities

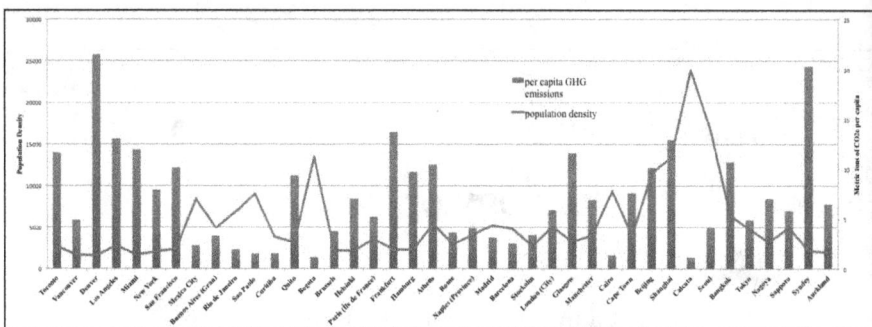

Source: author's compilation based on climate change action plans of selected cities and on UN-Habitat and World Bank databases.

Regarding *Urban Metabolism* and *Urban Political Ecology*

The unsustainable character of urban settlements is clearly visible when verifying its metabolic profiles, meaning, '…the process of contiguous de-territorialisation and re-territorialisation through *metabolic circulatory flows* organised through social and physical conduits' (Swyngedouw, in Swyngedouw et al 2005: 22).[1] Such analysis considers cities as systems open to energy and material flows, that is, those that take energy and materials from outside of the (urban) system and dispose depleted energy and materials inside but mainly outside of the system. See Figure 2 for an urban metabolism framework or the general urban inflows and outflows patterns.

Figure 6.2: Urban Metabolism Framework – Urban Biophysical Flows –

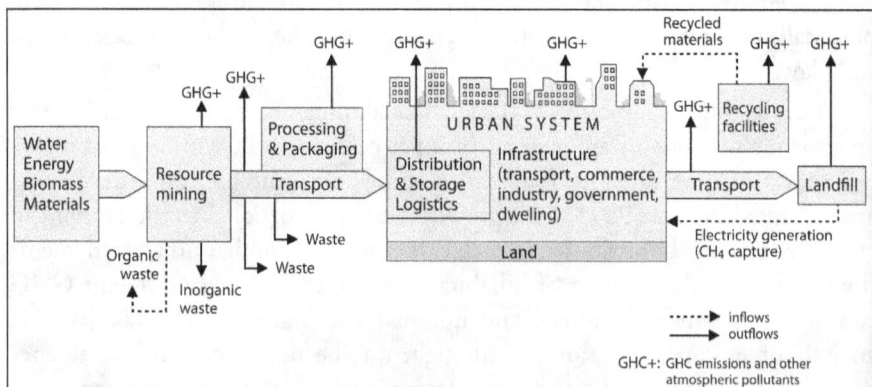

Source: author's own elaboration. Graphic design: Ángeles Alegre-Schettino.

Wolman's work is considered pioneering in empirical terms as it analyses energy and material flows into and out of a hypothetical 1960s United States-city populated by 1 million inhabitants. Wolman properly grasped the complexity and variability of said metabolic flows: such city required 625,000 tons of water daily and generated 500,000 tons of sewerage daily. In addition, 9,500 tons of fuel and 2,000 tons of food were required per day (Wolman 1965).

Diverse analyses of actual cities have been later completed, mostly in developed countries and focusing on various or specific metabolic flows (water, food, energy, etc.). The contributions made by Baccini and Bruner (1990 and 2012) as theoretical and methodological precursors, and later on by Kennedy et al. (2007, 2009 and 2011) and Minx et al. (2010), are notable because they provide a broad, integrated view of the evolution of urban metabolism research. In Latin America, the case of Bogota has been examined in detail (Díaz-Álvarez, 2011), while a generic comparative assessment has been carried out as well for the region's megacities and other capital cities (Delgado et al. 2012; and Delgado 2013).

Metabolic profile analyses of urban settlements and future projections allow, normatively speaking, to model more-or-less efficient routes for using resources and managing waste and, consequently, focusing efforts, for example, by planning metabolic dynamics starting from the design of infrastructure itself (or the so-called 'urban stock'), an objective that can be carried out through policies such as governmental incentives or even restrictions or coercive methods. Indeed, the metabolic challenge is to identify more efficient approaches and measures towards better-integrated human settlements, all with the purpose of minimising both, per capita and total consumption of energy and materials.[2] This certainly includes the prevailing need of reducing GHG emissions, which since 1970 have been generated up to 78 per cent due to burning fossil fuels and industrial processes – both intrinsically of urban nature (IPCC 2014).

This positive feature of urban metabolism as a potent analytical tool of the biophysical dimension of urban settlements must be, however, complemented with a socio-political analytical tool such as *urban political ecology* in order to be able to identify causes and processes leading to inequalities between rich and poor, or, in other words, the nature of space production that defines socio-political and biophysical conditions suitable for capital accumulation and thus uneven development (Harvey 1996).

Furthermore, it must be kept in mind that usually urban configurations are, at some point, outlined by land dispossession, grabbing, and speculation, followed – especially in a neoliberal context – by a much-more intense privatisation of the commons goods and state properties, including basic infrastructure for providing public services and amenities: this is, from water, sanitation, energy and transportation to green spaces. Hence, it can be said that cities are built to a significant degree under the impulses and needs of what Harvey (2004) has called *accumulation by dispossession*. It is a process that is possible with the support (or the 'absence') of the State, as it allows not only the segregation and gentrification of certain neighbourhoods, and the uneven erosion of public services and public space, but also a general loss of urban resilience and, therefore, an increase of vulnerability. This is, for example, an outcome of land speculation, irregular urbanisation and/or the loss of surrounding conservation space which has important ecological and climate functions for cities, such as: preservation of local biodiversity, water infiltration, carbon capture, cooling, among other so-called 'ecological services'. Yet, such negative outcomes mentioned above allow an appealing accumulation of capital, certainly based on a profoundly unequal model of space production that privatises benefits and socialises costs of all kinds, including those of socioecological nature. In this context, it is not by chance the increasing construction of gated communities; on the contrary, itis certainly a need of the middle and upper classes since they asymmetrically appropriate the positive aspects of life in the city.

Social resistances contending for a *right to the city* – a more equal, sustainable, inclusive, equitable, and supportive city – have given rise to the so-called *urban political ecology* since the revindication of said right involves the social right to manage the metabolic circulatory flows (Swyngedouw et al 2005).

As urban political ecology acknowledges, in current capitalist social relationships of production, '...the material conditions that comprise urban environments are controlled, manipulated and serve the interests of the elite at the expense of marginalised populations' (ibid, 6). Accordingly, and due to deep-rooted dominant power relationships, energy and material flows and stocks are unequally appropriated through market relationships or even straightforward dispossession.

The consequence of unequal purchase capacities is that the best constructions, highest-quality services, and the majority of increasingly privatised public space, are reserved for the 'best' consumers, that is, the upper and middle classes. Concurrently, the negative externalities of urban life tend to be exported as much as possible to poor-peripheral neighbourhoods or outside the city. Questions of class, race, ethnicity and so on are thus central to the process in terms of power relationship mobility capacity in order to define who gets access to, or control over, and who will be excluded from access to, or control over, natural resources and other components of the urban space (Swyngedouw et al 2005), including the imposition of the socio-environmental impacts that arise.

Urban Metabolism and the Political Ecology of Water: The Case of Mexico City and its Metropolitan Area

The case study that supports a better understanding and empirical evidence on the political ecology of urban metabolism comprises the *hydropolitan region*

Perló and González 2009) of the Mexico City Metropolitan Area (ZMVM, for ts acronym in Spanish), a territorial unit that Peña (2012) prefers to call the *city-basin*, which in this case refers to the interconnection of four basins that are not naturally related in any way: the Valley of Mexico, Alto Lerma, Cutzamala, and Tula basins. Such hydrosocial cycle (Swyngedouw et al. 2005), that is, the particular management of water in a given socio-environmental context has resulted in the appropriation/dispossession of water from neighbouring basins and the expulsion of sewerage towards another.

While the first three basins supply water inflows, the last is the destination of water outflows, including rainwater that cannot be infiltrated and that has historically led to the flooding of certain areas of the constructed space within the Valley of Mexico. The latter is due not only because of urban expansion itself, but also to the fact that the city is located over an endorheic system that has been progressively drained since the colonial period, specifically beginning with the construction of the Huehuetoca Royal Chanel (1607), the Nochistongo Tajo (1789), the first (1905) and second (1954) Tequixquiac tunnels, the deep drainage system (1975), and the more-recent Emisor Poniente deep-Tunnel (2010). Today, the system as a whole allows for the expulsion of some 57 m3/s.

It must be mentioned as well that, given to climate change, rains have become more intense over shorter periods, increasing water precipitation from 600mm in 1900 to 900mm at the end of the first decade of the 21st century (data form the Tacubaya meteorological station have registered a 7 per cent increase in water precipitation just since 1979). This situation, compounded with the climate change projection for 2050 regarding an increased amount and intensity of rainfall, especially during the rainy season (Aponte 2013), makes necessary to keep the water supply system, as well as tunnels, deep tunnels (*emisores*), and the deep drainage system that expels water outside the Valley, up to date as there is a direct relationship with the degree of the city's vulnerability to water availability (during dry seasons) and flooding (during rainy seasons).[3]

In spite of the above, the increase in the amount and intensity of rainfall by 2050 doesn't necessarily mean there will be sufficient water to cover the projected growing metropolitan demands (ibid). To better understand this condition it's necessary to analyse current water urban metabolism dynamics.

Inflows

As seen in Figure 3, Mexico City Metropolitan Area water inflows mainly come from:

- more than 600 wells that extract water from the Valley of Mexico aquifer (approximately 59 m3/s) which is currently being overexploited at a rate of up to a 1 m drop per year in the static water level (with a deficit of approximately 28 m3/s)[4];

   - the Lerma and Cutzamala basins system (approximately 5 m3/s and 15 m3/s, respectively); and
   - urban rivers and springs (approximately 2 m3/s) (Burns, 2009).

All together supply the bulk of water which is distributed by two administration entities: in the Federal District by the Mexico City's Water System (SACMEX) and throughout the State of Mexico by the State of Mexico Water Commission (CAEM) which delivers water to the corresponding metropolitan municipalities

Figure 6.3: Urban Water Metabolism of Mexico City Metropolitan Area (ZMVM)

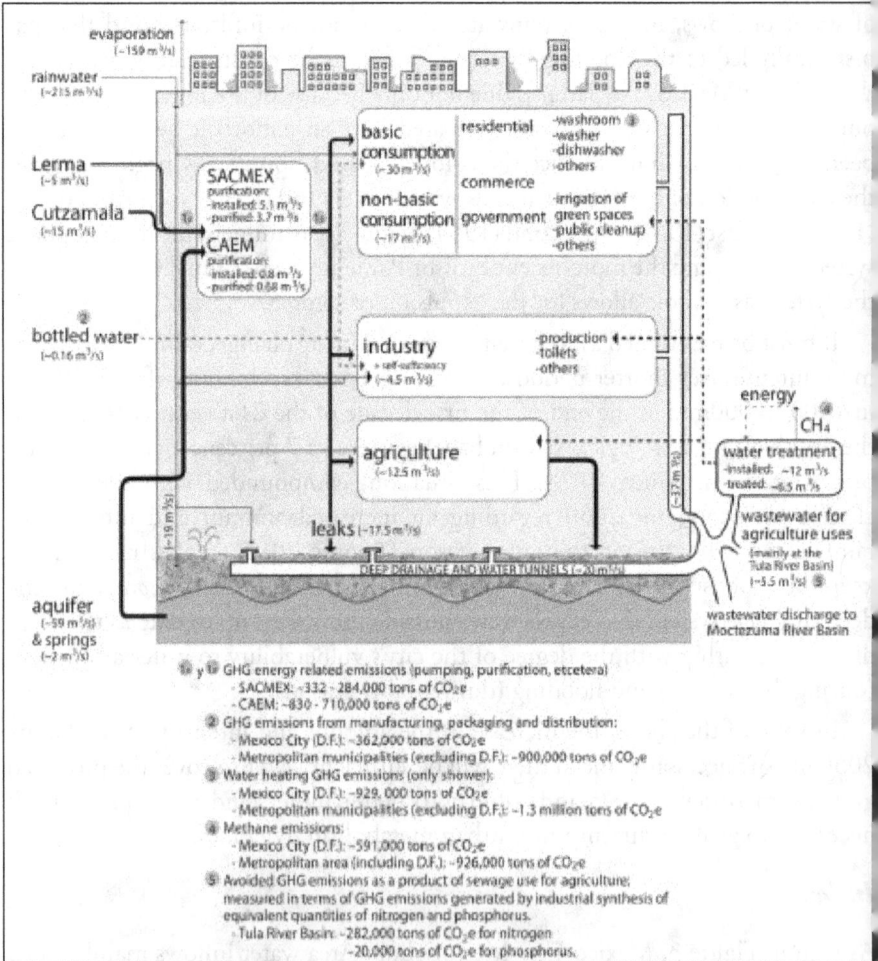

Source: Author's own elaboration based on Burns, 2009; Delgado, 2014B; SM-DF, 2012; SEMARNAT/CONAGUA, 2012; and date from INFO-DF, the local access public information entity.

Others minor sources include water self-supply systems, irregular water-truck delivery service (independent of the above government entities), and clandestine wells whose exact number is unknown but it is estimated at around 2,250 (Peña 2012).

In addition, the Mexico City Metropolitan Area imports bottled drinking water from several places, the bulk of which are domestic, and the remainder foreign. Bottled water consumption by the Federal District has been estimated at 2.07 hm3/year, and that of the metropolitan municipalities officially part of the states of Mexico and Hidalgo, has been estimated at 3.1 hm3/year. Together, this represents an inflow of approximately 0.16 m3/s, however, total demand for bottled water in the metropolitan area (ZMVM) has been calculated at 8.78 hm3/year when including the additional water necessary for its production or what is called *virtual water*.[6] At this point, it should be noticed that 76.94 per cent of the Federal District's population consume bottled water, while only 10.84 per cent boil it; 4.37 per cent filter or purify water by using other methods, and 4.58 per cent consume it directly from the tap (Jiménez et al. 2011). Similar patterns of consumption are seen in the rest of the metropolitan area.

Without including bottled water, water consumption in the Federal District averages some 327 litres per capita/daily; however, losses of between 35 per cent and 40 per cent due to leakage must be subtracted from this number (Jiménez et al. 2011; Peña 2012). In environmental terms, this is not a minor issue, especially when taking into consideration that about one-third of the total water consumed by the metropolitan area coming from the Lerma-Cutzamala system, must be pumped 1,100 m. The energy used for this purpose represents 80 per cent of the system's operating cost (Aponte 2013). Moreover, in the case of the Lerma system, installed capacity has been reduced from 15 m3/s to approximately 5 m3/s due to land-subsidence registered throughout the system as a result of over-pumping aquifers of the region. This has also contributed to the increase in the cost of pumping water from current low areas of the system to high areas.

Besides all difficulties and the economic and environmental costs just mentioned, it is obvious that water availability is greatly asymmetrical: distribution ranges from 177 litres in Tláhuac Borough to 525 litres in Cuajimalpa Borough. Boroughs with the highest incomes fall within the consumption range of 400 litres to 525 litres per capita daily (based on Jiménez et al. 2011).[7]

In addition to such water access inequalities in terms of quantity, there are as well disparities in terms of its quality (Jiménez et al. 2011; Díaz-Santos 2012); a reality that echoes SACMEX's purification capacity limitations: the entity has 38 purification plants in operation with an installed capacity of 5.1 m3/s but an actual purified flow rate of only 3.7 m3/s (SEMARNAT/CONAGUA, 2012; INEGI, 2014). Three additional plants (two in Chimalhuacan and one in Tlamanalco) corresponding to Mexico State municipalities belonging to Mexico City Metropolitan Area must be added: these plants have a total installed capacity of 0.8 m3/s and an actual flow rate of 0.68 m3/s (SEMARNAT/CONAGUA, 2012).

Outflows

Metropolitan water outflows (see Figure 3) have been estimated at an average volume of 57 m3/s, most of which are not treated.

Treatment installed capacity in Mexico City is of about 6.7 m3/s, with an actual treated flow rate of only 3 m3/s, while in the municipalities of the State of Mexico belonging to the metropolitan area, the installed capacity is of about 5.1 m3/s with an actual treated flow rate of 3.6 m3/s (based on SEMARNAT/CONAGUA, 2012). The remainder wastewater and stormwater is conducted to the Tula Basin (Tula-Moctezuma-Pánuco River) via the aforementioned deep drainage system and the Grand Canal. Up to 60 per cent of such flow may be treated at the Atotonilco treatment plant in the state of Hidalgo, one of the largest treatment facilities in the world currently under construction.[8]

Most of the treated water in the Federal District is used in urban green spaces (83 per cent), and the remainder is reused by industry (10 per cent) or in producing food in peripheral urban areas (5 per cent). Treated water in the State of Mexico's municipalities is conversely used for agricultural activities.

Notice that water treatment process becomes a key matter for two central reasons. The first is due to environmental and sanitary reasons, and, second, because, in principle, it allows the private sector to define who receives treated water and who doesn't (unless regulations indicate otherwise). Peña (2012) has already warned about the chance of such *de facto* privatisation of treated water.

Socio-environmental Jjustice Movements and the Urban Political Ecology of Water

In addition to the foregoing, water political ecology in the Valley of Mexico is particularly intense due to limited availability of water in the face of a strong demand (mainly because of a very dynamic urbanisation process in previous decades).

It is no coincidence, therefore, that a review carried out nationwide between 1990 and 2002 of some 5,000 newspaper articles on water conflicts, found that 49 per cent of such conflicts took place in the Valley of Mexico (Jiménez et al. 2011). Social mobilisations included public demonstrations and facility takeovers. About 56 per cent were due to a lack of water and 24 per cent to a hike in prices. In the metropolitan area, the districts that experienced the most social unrest were precisely those with less access to water due to a lack of sufficient infrastructure, such as certain areas of the east of Mexico City and the conurbation (e.g. Cerro de la Estrella in Iztapalapa Borough) (ibid).

In addition to such type of conflicts, there are others in areas where water is captured (urban water inflows) and expelled (urban water outflows). The former involves records from the rural mobilisation in the 1970s against the construction of the Lerma system in the Toluca Valley because it signified the potential loss of harvests due to a lack of water that instead was going to be sent to Mexico City Metropolitan Area. This conflict was however 'solved' with the payment of damages, at first using corn and later money, but the damage of the water system regime and the loss of productive agricultural land – along with an intense installation of industrial parks – soon revealed the socio-ecological impacts of such water transfer system.

In 1990, another conflict arose in the Temascaltepec area, where an expansion of the Cutzamala system was planned in order to obtain an additional 5 m3/s; however the project was halted due to social mobilisation (Jiménez et al. 2011).

By 2003, the improper operation of Villa Victoria Dam, part of the Cutzamala system, flooded 300 hectares of cropland, an event that led to a rural mobilisation seeking to obtain an economic compensation. Due to a lack of government response, the social movement revindicated its indigenous identity, shortly emerging as the so-called Mazahua Women's Army in Defence of Water (*Ejército de Mujeres Mazahuas por la Defensa del Agua*). Once it gained public attention with such striking pacific social movement, the public was able to learn about the lack of water in Mazahuas communities due to the appropriation of large amounts for supplying Mexico City Metropolitan Area.[9]

Further conflicts can be mentioned, the most recent case being the May 2014 struggle in San Bartolo Ameyalco due to the intention of piping water from a local spring to transfer it to the Cutzamala system.

As already mentioned, other water battles relate to the 'wastewater usufruct', meaning wastewater capture for treatment and the ensuing takeover of a large portion traditionally used for crop-production, mainly by rural inhabitants which, in our case study, mostly applies to the Mezquital Valley in the State of Hidalgo. This is a region with the lowest national levels of rainfall (400 mm/ year), yet some 85,000 to 90,000 hectares are cultivated and irrigated by using wastewater directly, which in turn provides about 44,000 tons of nitrogen and 17,000 tons of phosphorus (Burns 2009).

In 2011 the Irrigation Water of Agricultural and Cattle Producers Union (*Unión Productora Agrícola y Ganadera de las Aguas para Riego*) denounced, once again, a wastewater flow reduction (up to two-thirds) and hence the loss of grown vegetables and alfalfa. Such episode gave rise to proclamations such as the 'wastewater is ours!' (*Jornada, La.* 2011), reflecting the intensification of local conflicts between upland and lowland farmers in dispute for the 'white waters' from the Requena and Endhó dams and the 'black waters' discharged from Mexico City Metropolitan Area. In 2013 some 75,000 producers again protested against the lack of wastewater (Montoya 2013).

Furthermore, concerned about the installation of the Atotonilco treatment plant, the Council of Users in Defence of the Wastewater (*Consejo de Usuarios en Defensa de las Aguas Negras*) have already requested the local Congress of Hidalgo to ask the National Water Commission (CONAGUA) for an official document that could guarantee an equal volume of free wastewater for its croplands in relation to the volume currently being used given the possibility that treated water could attract charges (*NewsHidalgo* 2012).

In short, the complex conflict over Mexico City Metropolitan Area wastewater is certainly ongoing as well as its socio-environmental implications.

Water-Energy Nexus: A Climate Change Challenge

In 2013, Mexico City's Water System (SACMEX) used a total of 570.98 million kWh of which, among other things, 715,141.8 billion m3 of the total 953,522 billion m3 of distributed water was pumped (the remaining volume was moved using gravity) (INFODF 2014). As indicated in Figure 3, water inflows related emissions per cubic meter of water managed by SACMEX that same year reached 0.349 g - 0.298 g of CO_2, for an annual total of about 332,000 to 284,000 tons of CO_2.[10] Estimations for the case of conurbation municipalities served by CAEM, suggest a range of about 830,000 to 710,000 additional tons of CO2e.[11]

Total metropolitan water outflows related emissions, in terms of methane emissions, have been estimated at ~1.5 million tons of CO2e, of which 591,000 tons of CO2e correspond to the Federal District (SMA-DF, 2012).

End-user emissions, among which residential emissions play the greatest role, and must be added to the above. Such energy use corresponds to water heating/cooling processes. In the case of Mexico, the Ministry of Energy (SENER 2013) considers that only water heating stands for at least 13 per cent of the total amount of energy consumed in this sector while representing the third-highest home expense.

Heating water energy consumption in Mexico City Metropolitan Area has been estimated for 2006 at 31.2 petajoules, or 46 per cent of the total amount of energy consumed in the residential sector. Related emissions were in the order of 1,949,224 tons of CO2e (SMA-DF, 2008: 45). Data in this regard for 2010 indicated energy consumption at 33 PJ, that is, an annual per capita consumption of 1,654 GJ, with average per capita annual emissions at 105 kg of CO2e (SMA-DF, 2012).[12] If the current population of 22 million people in the metropolitan area is taken into consideration, emissions from heating water round to 2.3 million t of CO_2.

Meanwhile, emissions related to the consumption of bottled water, including manufacturing, packing, and distribution, reached 362,400 tons of CO2e per year at the city level and 900,900 tons of CO2e per year at the metropolitan level.[13] In addition, plastic waste generated by the consumption of bottled water reached 80,351 tons at the city level and 199,742.4 tons at the metropolitan level.[14] This s waste that must be collected and recycled or disposed of, as the case may be, which demands significant energy consumption that could be decreased or even avoided (in addition to the environmental impact derived from the disposal into the environment of increasing amounts of plastic) if tap water was treated to safe-drinking standards or if other drinking water options, such as public water fountains, were available.

Likewise, avoided emissions due to the use of 'black water' through a process that recovers nutrients (phosphorus and nitrogen) from water and that, therefore, are no longer needed to be industrially synthesised, have been estimated at some 414,500 tons of CO2e for both, the cropland area of Tula and of the Valley of Mexico.[15] In this context, it must be pointed out that although there are methods for safely using wastewater (Duncan and Cairncross 1989), these certainly were not implemented in the case under review.[16]

Taking into consideration that total GHG emissions from the metropolitan water-energy nexus described above reach some 5.5 million tons of CO2e annually (or 10 per cent of total metropolitan GHG emissions estimated for the year 2010)[17], it seems evident that an integrated water management planning should include all the complexity of urban water metabolism dynamics in order to generate smarter mitigation solutions at diverse space scales and timeframes, but also for achieving better (water) systemic outcomes, synergies and co-benefits of diverse kind, including those of social justice.

Co-benefits can be particularly significant in terms of electricity generation from methane capture and thus for climate mitigation; for land and water conservation through reducing plastic pollution and thus bottled water consumption (but also by effectively protecting conservation land and promoting reforestation programmes); for avoiding GHG emissions by properly using wastewater for crop production (meaning taking the necessary precautions for protecting health and the environment); among others related, for instance, to public health improvement.

Final Remarks: A Focus on New Paradigms

The capacity to transform urban settlements in developed countries is incomparably greater than that of developing ones, not only because they have greater means of economic and technological innovation, but also because many hidden or indirect socio-environmental and climate costs are usually 'exported' or internationalised (in spite of the fact that all cities do this in one way or another using their own hinterland).

Therefore, urbanisation in developing countries tends to be more problematic and complex due to a limited or overburdened capacity to take measures and actions, a scenario in which urban poverty is an enormous further challenge to any type of more human and sustainable urban reconfiguration.

The design and execution of public policies needed for transforming the current trend of constructing, operating, managing, and living in cities must be proactive, imaginative, and based on an integral metabolic planning. Accordingly urban metabolic analyses seem to be essential to policy tools, but in general urban planning (which includes land planning), can be adjusted to overarching contextual changes and to historical trends and socially desirable futures.

The sum of multiple actions, provided these actions commence with the aforementioned integrated planning with social justice, may have a heightened impact while allowing synergies and co-benefits of various types. Consequently traditional sector-based management is no longer sufficient or viable.

Even more, a profound transformation of existing urban settlements demands not only considering the urban *form* design and its metabolic profile, but also a profound reformulation of the urban territorial *function* or the purposes of urban territorial configuration; this means moving from schemes intending capital accumulation as a priority, to those that promote human development or well-living instead. In that sense, and as Swyngedouw et al (2005) correctly claim, it's then crucial to ask who produces what type of urban configurations, who gains and who loses and in what ways, and who benefits and who suffers from particular urban territorial configurations (within and beyond the cities).

Likewise, given that the *construction of space* is dynamic, it is equally important to understand which are future perspectives, and based on which cultural, historical, and environmental notions this or that approach provides, or does not provide, alternatives (as well as alternatives to what).

We are currently in a position in which not only technological solutions regarding the type and design of infrastructure play a part, but also in which a deep-seated change in prevailing logic and, therefore, the nature and desirability of the solutions themselves, is crucial.

In the specific case of the hydrosocial cycle of Mexico City Metropolitan Area, the challenge is of great significance, though there are certain factors that are already evident. In this regard, in spite of the fact that the population is growing slowly in the area that is already urbanised (there has been a significant increase in some municipalities of the conurbation), there is a confirmed need for 1 m3/s of additional water by 2015 alone. This represents a demand being managed against a backdrop of a decreasing flow rate in supply of 3 m3/s over the past decade. Therefore, in the coming years, it is clear that the highest pressure will be seen in the conurbation area, though changes in social expectations regarding the amount and, especially the quality of water, could also pose significant challenges for the supply system (Jiménez et al. 2011).

In recognition that the availability of water has already reached its maximum viable point in spite of all technological innovations and big infrastructure development, a projected solution, not free from social dispute, is committed to expand the water system's into more-remote regions such as in the case of the fourth phase of the Cutzamala system aimed at making use of the high-altitude Tula, Tecolutla, Valle del Mezquital, and Amacuzac rivers, which in most cases will yield comparatively lesser volumes of lower-quality treated water.

A long-term solution cannot be that simple, nor can it be merely or mainly centred on large-scale engineering solutions. The so-called 'new water culture' centred around a moderate and responsible, though socially fair consumption, is certainly an important issue, though addressing the issue of leaks in present and future systems; attempting to effectively plan the use of land, especially peripheral urban land; protecting conservation land while restoring its vegetation as well as that of the city itself for the purpose of, among others, increasing evapotranspiration capacity and reducing city temperature; are all extremely important issues as well.

Given the heterogeneity of water conditions and territorial infrastructure that hinder the ideal homogeneity of the equal supply of water in terms of quantity, quality, and frequency throughout the entire Valley of Mexico, it is also desirable to decentralise the water system by adding to multiple spatial scales other systems of less import that might increase the flexibility, transformation, and resilience of the whole system in the face of external shocks, including those resulting from climate change (e.g., rainwater harvesting, local water reuse/treatment, and so on).

In summary, new paradigms for managing water that are more fair from a social and ecological perspective and that are more harmonious in the short, medium, and longterm, demand a combination of new technologies, practices (including planning and regulation), and values that must be developed and implemented by all inhabitants, including the social, political, and economic players of each territory. This process is feasible, though certainly slow due to both, the 'infrastructure lock-in' (to certain metabolic dynamics and patterns) and to the persistent nature of practices and interests grounded in traditional management criteria.

A genuine bottom-up management of water infrastructure (and certainly of the urban built environment as a whole), which goes further than just citizens' participation is and will be of more importance in order to formulate new manners of democratic self-management with a sense of community. This is a scenario of genuine participative democracy, which, from a broader perspective, must be viewed as a mechanism for empowering the people inhabiting the same territories for the purpose of, among other matters, guaranteeing human rights, such as the right to water and sanitation or a healthy environment, but more over for supporting people's right to a fair city.

Acknowledgments

Paper presented on 24 July 2014 at the Comparative Research Workshop on 'Inequality and Climate Change: Perspectives from the South' in Dakar, Senegal; an activity backed by the Council for the Development of Social Science and Research in Africa, the International Development Economics Associates and the Latin American Council of Social Sciences. Special thanks to the Journal of Political Ecology editors for granting permission to publish this paper as a book chapter as it was originally published in vol. 22 of the above mentioned journal.

Notes

1. For Swyngedouw (Swyngedouw et al. 2005: 22 and 25), 'metabolic circulation' refers to the merging of existing (bio)physical dynamics with the set of conditions that regulate and frame current social relations of production in this or that territorial space.
2. It is notorious that 'green economy' policies only pay attention to per capita efficiency or at the level of subcomponents of the system. Since such relative efficiency does not guarantee an absolute efficiency, a rebound effect is a usual outcome.
3. Certain areas of the city consistently demonstrate flooding; including the overflow of wastewater in areas such as Valle de Chalco and Ixtapaluca caused by a lack of sanitation in the city's drainage system and by rains that increase in intensity. In 2010, overflows left almost 25,000 people homeless.
4. The aquifer is currently overexploited. The historic level of water extraction is estimated at 2 m3/s for 1870; approximately 22 m3/s for 1952 (which shows already a deficit, as the there were only 19 m3/s of recharge); and some 59 m3/s extracted by 2007 (Burns, 2009).
5. Mexico City Water System is an entity that articulates private concessionaries of water public service, geographically arranged in four operating areas: (1) Proactiva Media Ambiente SAPSA of the Mexican ICA and the French Veolia; (2) Industrias del Agua de la Ciudad de México S.A. de C.V. of the Mexican Peñoles and the French Suez; (3) Tecnología y Servicios de Agua S.A. de C.V., which is also property of Peñoles and Suez; and (4) Agua de México, S.A. de C.V., of national capital.
6. Estimate based on 2009 national per capita consumption of 235 l/year (Delgado, 2014B). Average indirect water demanded for each litre of bottled water, or virtual water, has been estimated at 700 additional millilitres according to figures provided by FEMSA Coca-Cola and Nestlé Mexico in 2010.
7. About 38.4 per cent of the Mexico City's population receive water only a few hours per day, while 61.5 per cent receive it all day. Yet, on average, 52-53 per cent of the poorest and moderately poor areas of the city receive water only a few hours per day. This figure is only 18-19 per cent in the least-poor areas where in addition most of inhabitants have economic means to invest on proper dwelling water tanks and pumping infrastructure (Jiménez et al. 2011)
8. Operated by DEAL, a company owned by Carlos Slim, is supposedly capable of treating 60 per cent of the metropolitan sewerage at an energy cost of 166 million

kWh/year and an average generation of 917 t of sludge daily. The plant is expected to come into operation in February 2015. It's located in the ejido 'Conejos', right in the middle of a road that communicates San Antonio and San José neighbourhoods, both lacking of basic water and sanitation services.

9. The Mazahuas movement also opposed the then new National Water Act, which excluded community water management systems in indigenous territories. In spite of being criminalised, this social struggle still continues (Ávila-García 2011).

10. Calculation based on two different methodologies used for estimating electricity production-related GHG emissions in Mexico: 1) the National Commission on Efficient Energy Use approach which estimates emissions at 0.5827 tons of CO2e/MWh based on electric power consumption, and 2) the Programa México approach that estimates emissions at 0.498 t of CO2e/MWh based on electrical power generation data (SEMARNAT/INECC, 2012: 42).

11. Calculation derived from SACMEX's estimate, assuming the same emission factors per cubic meter, and considering that CAEM's users sum about 60 per cent of the total users Mexico City Metropolitan Area.

12. The calculation for both years is based on four showers per week per inhabitant of Mexico City Metropolitan Area using 45 litres per shower (SMA-DF, 2008; SMA-DF, 2012). Said calculation depends on one person's habits: taking six showers per week, using 65 litres of hot water, will instead generate 208 kg of CO2e per year (SMA-DF 2012).

13. The emissions estimate is based on a factor of 175g per litre of bottled water according to information provided by Nestlé-Mexico (Delgado, 2014B).

14. Plastic waste generation was estimated at 38.8g (bottle plus secondary packing) per liter of bottled water according to information provided by Nestlé México (Delgado, 2014B).

15. Wastewater contains 97 per cent water and 3 per cent solid materials (organic and inorganic). The following are the factors used in calculating emissions: 6.41 kg of CO2e per kg of nitrogen and 1.18 kg of CO2e per kg of phosphorus (based on estimates by García et al. 2011).

16. For instance, it is well known that treatment processes are indeed required to remove toxic substances and pathogens before any irrigation can take place. Nonetheless, as already described, treatment capacity of Mexico City Metropolitan Area is still extremely low.

17. Total 2010 emissions for Mexico City Metropolitan Area were 54,700 Gg of CO2e (SMA-DF 2012).

References

Angel S., J. Parent, D.L. Civco, A. Blei, and D. Potere, 2011, 'The Dimensions of Global Urban Expansion: Estimates and Projections for all Countries, 2000-2050', *Progress in Planning*. 75. Pp. 53–107.

Aponte Hernández, Nidya Olivia, 2013, 'Metodología para evaluar la disponibilidad del agua para uso muncipal y sus costos bajo los escenarios de cambio climático'. Programa de Posgrado en Ingeniería Ambiental. Posgrado de Ingeniería, UNAM. Mexico.

Ávila-García, Patricia, 2011, 'Water Conflicts and Human Rights in Indigenous Territories of Latin America', Rosenberg International Forum on Water Policy. Available online at: http://ciwr.ucanr.edu/files/168776.pdf. Accessed on 29 March 2014.

Baccini, P., and Brunner, P., 1990, *Metabolism of the Anthroposphere*. First Edition, Cambridge, MA. / London: MIT Press.

Baccini, P., and Brunner, P., 2012, *Metabolism of the Anthroposphere*. Second Edition, Cambridge, MA. / London: MIT Press.

Burns, Elena, coordinadora, 2009. *Repensar la Cuenca: la gestión de los ciclos del agua en el Valle de México*. UAM-X/USAID. Mexico.

CONAGUA, 2014, 'Planta de tratamiento de aguas residuales Atotonilco', on line [03/29/14]:www.conagua.gob.mx/sustentabilidadhidricadelValledeMexico/ProyectoDrenajes.aspx?Pag=3

Delgado-Ramos, Gian Carlo, 2013, 'Climate change and metabolic dynamics in Latin American major cities' in: Zubir, S.S. y Brebbia, C.A., editors, *Sustainable City VIII. Urban Regeneration and Sustainability*, Southampton: WIT Press, pp. 39 -56.

Delgado-Ramos, Gian Carlo, 2014a, 'Ciudad, agua y cambio climático. Una aproximación desde el metabolismo urbano'. *Medio Ambiente y Urbanización*. Vol. 80. No. 1. Instituto Internacional de Medio Ambiente y Desarrollo América Latina. Buenos Aires, Argentina. Pp. 95-123.

Delgado-Ramos, Gian Carlo, 2014b, 'El negocio del agua embotellada'. *Realidad Económica*. No. 281. Instituto Argentina para el Desarrollo Económico. Buenos Aires, Argentina, 1 January – 15 February.

Delgado-Ramos, Gian Carlo., Campos Chávez, Cristina., and Rentería Juárez, Patricia, 2012, 'Cambio climático y el metabolismo urbano de las megaurbes latinoamericanas'. *Hábitat Sustentable*, Vol. 2, No. 1. Santiago de Chile, Chile. Pp. 2 – 25.

Díaz Álvarez, C.J., 2011, *Metabolismo de la Ciudad de Bogotá: una herramienta para el análisis de la sostenibilidad ambiental urbana*. Universidad Nacional de Colombia. Bogota, Colombia.

Díaz-Santos, María Gudalupe, 2012, *Implicaciones sociales de los contratos al sector privado en el servicio de agua potable en la Ciudad de México*. Bachelor Thesis, Faculty of Political and Social Sciences, UNAM. Mexico.

Duncan, Mara and Cairncross, Sandy, 1989, *Guidelines for the safe use of wastewater and excreta in agriculture and aquaculture*. Food and Agriculture Organization. On line: http://whqlibdoc.who.int/publications/1989/9241542489.pdf

García, Carlos., Fuentes, Alfredo., Hennecke, Anna., Riegelhaupt, Enrique., Manzini, Fabio y Masera, Omar, 2011, 'Life-Cycle Greenhouse Gas Emissions And Energy Balances Of Sugarcane Ethanol Production in Mexico'. *Applied Energy*, Vol. 88, No. 6, pp. 2088-2097

Harvey, David, 1996, *Justice, Nature and Geography of Difference*, Oxford, UK.: Blackwell.

Harvey, David, 2004, *El Nuevo Imperialismo*, Spain: Akal Ediciones.

INFO-DF, 2014, Dirección ejecutiva de operación. Dirección de Agua Potable y Potabilización. Sistema de Aguas de la Ciudad de México. Ofico de solicitud pública: 0324000009414, Mexico, 6 February.

IPCC – Intergovernmental Panel on Climate Change, 2014, *Climate Change 2014: Mitigation of Climate Change*. Chapter 12. AR5 – WGI. WMO / UNEP. Geneva, Switzerland. Available on line [5/15/2014]: http://mitigation2014.org/report/final-draft/

Jiménez Cisneros, Blanca., Gutiérrez Rivas, Rodrigo., Marañón Pimentel, Boris., and González Reynoso, Arsenio, coord., 2011, *Evaluación de la política de Acceso al agua potable en el Distrito Federal*. PUEC-UNAM. Mexico.

Jornada, La, 2011, 'Escasez de aguas negras incuba conflicto en Valle del Mezquital'. *La Jornada*, Mexico, 29 May.

Kennedy, C., Cuddihy, J., and Engel-Yan, J., 2007, 'The Changing Metabolism of Cities', *Journal of Industrial Ecology*, Vol. 11, No. 2, pp. 43-59.

Kennedy, C., Steinberg, J., Gasson, B., Hansen, Yvone., Hillman, Timothy., Havránek, M., Pataki, D., Phdungsilp, A., Ramaswami, Anu and Villalba Méndez, G., 2009, 'Methodology for Inventorying Greenhouse Gas Emissions from Global Cities. *Energy Policy*, Vol. 38, pp. 4828-4837.

Kennedy, C., Pinceti, S., and Bunje, P., 2011, 'The Study of Urban Metabolism and Its Applications to Urban Planning and Design.' *Environmental Pollution*, Vol. 159, Nos. 8-9, pp. 1965 – 1973.

McKinsey Global Institute, 2013, *Infrastructure Productivity: How to Save $1 Trillion a Year*. McKinsay. On-line [04/27/14]: www.mckinsey.com/~/media/McKinsey/dotcom/Insights%20and%20pubs/MGI/Research/Urbanization/Infrastructure%20productivity/MGI_Infrastructure_Full_report_Jan2013.ashx

Minx, Jan., Creutzig, Felix., Medinger, Verena., Ziegler, Tina., Owen, Anne., and Baiocchi, Giovanni, 2010, *Developing a Pragmatic Approach to Assess Urban Metabolism in Europe. A report to the European Environment Agency*. Stockholm: Environment Institute / Technical University of Berlin.

Montoya, Juan Ricardo, 2013, 'Acusan de despojo de aguas negras a Edomex y DF'. *MagazineMx*. Mexico, June 17th. On line [03/29/14]: www.magazinemx.com/bj/articulos/articulos.php?art=15157

News Hidalgo, 2012, 'Temen campesinos de Hidalgo, fatla de agua para sus cultivos'. News Hidalgo. Mexico, 12 June,. Available online at: www.newshidalgo.com.mx/?p=4925. Accessed on 29 March 2014.

Peña-Ramírez, Jaime, 2012, *Crisis del agua en Monterrey, Guadalajara, San Luis Potosí, León y la Ciudad de México (1950 – 2010)*. PUEC-UNAM. Mexico.

Perló, Manuel and González, Arsenio, 2009, *¿Guerra por el agua en el Valle de México? Estudio sobre las relaciones hidráulicas entre el Distrito Federal y el Estado de México*. PUEC-UNAM. Mexico.

SEMARNAT/CONAGUA, 2012, *Inventario Nacional de Plantas Municipales de Potabilización y de Tratamiento de Aguas Residuales en Operación*. Mexico.

SEMARNAT/INECC, 2012, *México – Quinta comunicación nacional ante la Convención Marco de las Naciones Unidas sobre el Cambio Climático*, México. Available online at: www2.inecc.gob.mx/publicaciones/download/685.pdf. accessed on 29 March 2014.

SMA-DF – Secretaría de Medio Ambiente del Distrito Federa,l. 2008, *Inventario de Emisiones de la Zona Metropolitana del Valle de México, 2006*. Gobierno del Distrito Federal, Mexico.

SMA-DF – Secretaría de Medio Ambiente del Distrito Federal, 2012, *Inventario de Emisiones de la Zona Metropolitana del Valle de México, 2010*. Gobierno del Distrito Federal. Mexico.

SENER – Secretaría de Energía, 2013, 'Calentador de Agua'. Comisión Nacional para el Uso Eficiente de la Energía. Mexico. Available online at: www.conuee.gob.mx/wb/CONAE/calentadores_de_agua. Accessed on 25 May 2014.

Swyngedouw, E., Heynen, N., and Kaika, M., 2005, 'Urban Political Ecology – Politicising the Production of Urban Natures' in: *In the Nature of Cities – Urban Political Ecology and the Politics of Urban Metabolism*, London, UK: Routledge..

United Nations, 2011, *World Urbanization Prospects, the 2011 Revision*. The Population Division of the Department of Economic and Social Affairs of the United Nations New York UN.

Wolman, A., 1965, 'The Metabolism of Cities', *Scientific American*, Vol. 213, No. 3, pp. 179-190.

7

Indigenous People and Climate Change: Causes of Flooding in the Bolivian Amazon and Consequences for the Indigenous Population

Gabriela Canedo Vásquez

Introduction

This chapter addresses the causes and consequences of climate change among an indigenous population. The effects of climate change on the indigenous people of Beni leading to the historical flooding that occurred there in 2014 was devastating. to the chapter focuses on the effects of the rains on Beni, as this is one of the places that suffer 'monsoon-like' attacks every year. These attacks result in frequent rains and main rivers breaking their banks, resulting in extreme flooding. While this area is characterised by frequent flooding, recent reports indicate that the growth of the rivers in 2014 was higher than those recorded in 2007 and 2008, years in which the effects of La Niña were present. This leads to increased loss of life and property for indigenous and peasant families in the region. On 4 February 2014, a hydrological red alert was declared for the Mamore basin affecting several municipalities in Beni. In this community alone there would have been 3,957 affected families, in addition to a 140-acre loss of crop and 218 families left homeless by the flooding.[1]

The main causes of flooding are climate change, deforestation and the overflow of the Brazilian dams, Jirao and San Antonio. Deforestation is one of the causes of excessive rains. Beni is one of the areas where livestock farming is the most important activity, which goes hand-in-hand with the cutting down of trees to enable grazing. Livestock is managed by private corporate cattle ranchers. However the harshest consequences are suffered by the indigenous people, who live on hunting and subsistence farming.

This chapter examines the effect of climate change on the indigenous people from an anthropological perspective. It analyses how land tenure and management in the area fall under two different main activities, livestock farming and agriculture, and how they are developed by two different social groups, namely, cattle ranchers and indigenous people. These activities point to the existence of two different social groups in the community as well as to their different occupations within the same geographical space. For cattle ranchers, the land is there to take advantage of for the purpose of developing their economic potential in the market, so deforestation and *chaqueo*[2] are common practices in order to rear livestock. Meanwhile, for the indigenous people, preservation of the territory has a more intrinsic worth and a broad understanding of this allows for their subsistence and the development of their community life, as well as managing resources collectively. Thus, these two groups are in constant conflict over land tenure and land ownership rights.

The magnitude of the last flooding occurrence compels a reflection on the implications of the management of the natural resources of this locality. Deforestation (as a more localised cause) and the increase in global temperature (as a more general cause) make it imperative to discuss climate change and its impact on the most vulnerable populations such as the indigenous one under consideration. The Bolivian Amazon, and specifically Beni, is an area of my own interest where I conducted my research, and where floodings have occurred rather frequently. This leads to the conclusion that Beni is one of the places where the indigenous population lives precariously, and where they are set to live through the consequences of climate change.

It is important to consider the overall context of climate change and its effect with indigenous people. Afterwards we will give an overview of Beni, highlighting the development of the economic activities of the farmers and the indigenous population and the implications of their operations on the land. Finally, the chapter examines the causes and consequences of the flooding and their impacts on the indigenous population. The methodology chosen is based on the well-considered view of:

> Studying and grasping the processes of disaster on a larger scale, but at the same time, identifying them with the contextual conditions and the specific threat, its manifestations, its effects, and impacts. Far from trying to get an overview, we address the specificity of the process of disaster risk and vulnerability, to which we have referred the process of disaster with a surname, the risk surname, and the vulnerability surname (García 2004)

Climate Change and the Indigenous People

It is well known that the earth's climate is changing radically from previous periods in history, and that the climate change is, in part, occurring due to the influence of human activity. According to the IPCC the global average surface temperature of the Earth has risen from 0.56 to 0.92 C during the twentieth century (Gallardo 2012: 21). And on average global temperatures will rise by between 1.1 and 6.4 C by the end of the twenty-first century (IPCC 2000; Mario Baudoin 2012:5).

Baudoin states that climate change is the product of global warming and will have significant social and economic impacts, especially in developing countries. It will produce significant changes in the frequency and intensity of extreme events such as heat waves, droughts, and floodings (2012: 1).

Facing the natural risks that could arise from climate change is a challenge for developing countries. In these countries, extreme climatic events tend to prolong development in the long-term because large amounts of money that could be aimed at improving the quality of life are invested in assisting the population and reconstruction after disasters (Baudion 2012: 5). Even if this is not the only cause of the failure of poverty reduction programmes, risks and disasters affect the results of these programmes (IPCC 2001). Climate change will have social and economic impacts on impoverished populations, and affect their health and food security.

In recent decades, Bolivia has suffered extreme weather events such as floodings, severe droughts, forest fires, and loss of water reservoirs that impact on the people's survival and responsiveness, and have exacerbated existing vulnerabilities. A 2011 UNDP study extracted data from EM-DAT[3] (1900 to 2010) showing the ten worst weather disasters that have taken place in the past three decades in Bolivia. The same database shows that droughts, floods, extreme temperatures, and catastrophic landslides have increased in frequency in recent years (UNDP 2011: 45). Records show that between 2000 and 2009, 911 people died and over 2,000,000 were affected by natural disasters. In lowland areas, extreme events such as storms, droughts, high winds, and flooding, led to significant economic losses, particularly in agricultural and livestock production. They also led to the damage of infrastructure and the emergence of infectious diseases (MMAA 2009; Githeko et al 2000 in Baudoin, 2012: 6).

The Indigenous People and their Vulnerability

The indigenous communities depend on natural resources for their subsistence and are, therefore, the most vulnerable to climate change. While people living in marginal lands have always been exposed to various types of environmental change and have developed strategies to address them, it is likely that this valuable knowledge of adaption is outweighed by the future risks that climate change will bring.

Using documents related to policies on climate change, including the Stern report and the Fourth Assessment Report of the IPCC 2007, (Oviedo 2008) asserts that those who will most suffer the consequences of climate change will be the poorest and most vulnerable communities in the world, including the indigenous and traditional people. Confronting climate change, the indigenous people are in a vulnerable position (defining vulnerability as the inability of the poor to manage risk). In general, the most vulnerable people face risky conditions to cover their basic needs which are highly sensitive to climate change (water, food, health) (UNDP 2011: 45).

Hampson distinguished in his report to the Commission on Human Rights 1998-2007, three levels in which indigenous people have their rights affected by environmental damage, and specifically climate change:

i) Damage that adversely affects their territories, such as biodiversity loss and water scarcity.

ii) Displacement – being forced off their land to other regions of the *same* country because they can no longer make a living in their territory. In addition to the physical loss of land, it may also involve losing their inalienable right to it especially if it is being colonised and no longer made consistent with their traditional habitat (condition is stipulated, in many Latin American constitutions, as a prerequisite for the right to own land).

iii) Displacement – being forced from their lands and fleeing to *another* country or location. It is feared that, in this case, they lose all their rights as an indigenous people, under which, according to the ILO Convention 169 they would no longer be indigenous, but an ethnic minority in a foreign country. Would they still be entitled to exercise their rights in their home country? What territories can the landless claim? What rights do they have? Under whose authority can they assert their rights? These questions are particularly topical for the people of the Caribbean islands threatened by floods (Feldt 2011: 5)

For these reasons, the indigenous peoples demand the recognition of their rights to land, consultation, participation even in the context of negotiations about climate; and that in the context of these negotiations their rights may prevail over other interests (Feldt 2011: 5).[4]

Oviedo (2008) notes that exposure to the impacts of climate change depends on where people choose or are forced to live. Indigenous people often live in difficult, fragile, and isolated places that make them vulnerable. Likewise, indigenous communities are dependent on materials and natural fibres to meet their needs for food, timber, fuel, generation of income, medicine, and for spiritual purposes. It is estimated that climate change will alter the availability and distribution of these resources. Similarly, floods and droughts have negative effects on crop production and also reduce the supply of fuel and water. Not to mention, climate change acts as a stressor on indigenous people and limits their ability to cope. Finally it is noted that climate change will further weaken the health of millions of people. Indigenous groups could be exposed to greater health risks related to water quality, or infections such as dengue and malaria.

The Amazon, Home to Indigenous Groups

One area that has received considerable attention has been the Amazon, mainly because it is one of the areas with the highest diversity of species on the planet. The impact of global change, however, is different within different ecosystems even within the Amazon itself. Bolivia is located on the southwest border of the Amazon, in a transition area to southern temperate ecosystems. Some estimations consider, for example, that this is precisely the region that will receive the greatest impact of global change, such as the shrinking of glaciers in the Andes (Francou et al 2005; Soruco et al 2008; Chevallier et al 2011).

Different predictive models of change in temperature and precipitation under different climate change scenarios indicate a trend towards loss of humidity in certain regions. The Amazon region is considered to be at particular risk, especially when one includes the impacts of increased land use. Regional models suggest a further increase of temperature in the Amazon because of substantial losses of the Amazonian rainforest and other vegetation that impact on the climate of the region due to reduced evaporation rates and changes in the precipitation regime (Cox et al 2000 in Baudoin 2012: 70) and because of global causes as the increased carbon dioxide and other gases such as nitrous oxide and methane (Alencar et al 2006 in Baudoin 2012: 70). Also, increased temperatures and decreased precipitation could increase the frequency of droughts in the Amazon, with negative effects on vegetation: the replacement of relatively fire resistant vegetation by other species that are highly flammable, producing more frequent and extensive fires.

Jong and Mery (2011) have pointed out the concern about the future of the Amazon region, the integrity of its forests, other ecosystems, and the welfare of rural populations, especially indigenous and other traditional peoples, because of the threats of climate change and food security problems.

Having briefly reviewed the relationship between climate change and indigenous people, we will now focus on Beni, where flooding had been strong and excessive in 2014.

Beni and the Amazon

Beni is one of the fragments of Bolivia which is part of the Amazon. It has an extraordinary exuberance; a variety of flora, fauna and social diversity (sixteen indigenous communities live in this territory). In 2014, Beni bore the brunt of the incoming rains, reporting a historical increase on inland precipitation.

Diversity is expressed, for example, in the entire plains of Beni, including the adjoining woodland which is home to about 5,000 plant species, and the savannahs themselves, containing 1500 of these species (Beck and Moraes 1997). In the floodplain of the central region of the Mamore River, the presence of about 900 species (Beck and Moraes 2004) has been reported. In summary, the plains of Beni show an interesting range of plant species and sub-formations, transforming it into a unique and valuable area (Baudoin 2012: 14).

Beni is characterised by a particular geomorphology with very little inclination, so it is subject to small annual flooding or large floods every few years that can even cover it almost entirely (Baudoin 2012: 11). This area, in particular, is regularly affected by seasonal floods, fires and droughts that have much to do with global climatic phenomena: El Niño and La Niña. During the wet season the water supply of the upper watershed of the Mamore and Beni, in addition to local rains, brings a build-up of rivers and flooding of a large proportion on the area. Moreover during the dry season, there are frequent wildfires caused mainly by the burning of grasslands, which are used as food for cattle to stimulate regrowth, and also to enable land for farming or livestock grazing.

There are complex social, economic, and environmental interactions that influence the vulnerability of the Beni population to cope and recover from natural disasters. Some determinants are: environmental degradation, poverty, and social inequality (ISDR 2001; Thomalla et al 2006; Oxfam 2010; Baudoin 2012: 8).

Social Diversity and the Presence of Indigenous People in Beni

Currently, Beni has a total population of 422,000 inhabitants. The indigenous population is around 105,000. Although it reduced indigenous population, 16 of the 36 indigenous groups of Bolivia live in el Beni.

Formerly, the population of Beni consisted of a large number of ethnic groups that were traditionally integrated into the scheme of the Jesuit reductions in the early seventeenth century. This historic event determined most of the current characteristics of the local population and the distribution of the main population clusters. Since the early nineteenth century a successive flow of immigrant traders and pioneers began to settle around the indigenous areas and were supported by unequal trade and trade-off relationships which began to concentrate resources by legal and illegal means. Also, due to issues related to the high mobility of the indigenous population, its territories were considered vacant land and gradually became the domain of cattle ranch owners (Urioste and Pacheco 2001: 105).

The process of penetration of livestock which was progressively moving indigenous communities to smaller and smaller areas, relinquished the indigenous of control of livestock production. This has defined major social, cultural, and economic contrasts. A set of conflicts between the indigenous population and the farm owners by the overlapping of property rights remains unresolved. The indigenous population is completely surrounded by cattle ranches.

Beni is an excellent region for rearing livestock with ranching characterised by the use of large amounts of natural pastures requiring little capital investment (Urioste and Pacheco 2001: 142). Meat supplies to the domestic market have determined the economic landscape of this region. Despite the favourable conditions mentioned above, the region has some disadvantages that threaten the stability of this production due to temporary and permanent flooding, as well as the limited availability of offshore areas for livestock protection when such natural phenomena occur (Urioste and Pacheco 2001: 105). For example, within the regional economic dynamics, a division of activities among social parties has appeared. For one thing, the doomed livestock farmers, indigenous people, and peasants have devoted themselves to the development of farming systems on the small-scale, thus allowing them to meet their subsistence needs.

We now focus on these two groups in order to see their interaction on the land.

Indigenous Groups and their Concept of Territory

Access to the property rights of indigenous territories in the Bolivian Amazon is irreversible. Positive trends can be observed showing that indigenous people and traditional communities have increased their territorial control from the 1990s, having expanded forest conservation and biodiversity efforts (Jong and Mery 2011: 5).

However, extreme poverty and neglect of this population will not only change property entitlement but also not guarantee sustainable use of forests and all the natural resources of flora and fauna in the *Tierras Comunitarias de Origen* (TCOs).

A general observation of the indigenous groups in the region's lowland is that there are less and less hunters and gatherers; instead there are increasing numbers of unstable farmers. That is, their lifestyle change and sedentary work leads them to live in increasingly stable settlements. Thus the indigenous families spend more time cultivating their farms and, more recently, also raising cattle (Urioste and Pacheco 2001: XXXII).

Their economic activity (subsistence through small-scale agriculture, hunting, and gathering) and their social cohesion as indigenous communities leads them to conceive of the geographical space they inhabit as a territory, a 'big house', 'the mother', and an 'unlimited' space where they may move freely, looking for the everyday food. Given these conceptions, the territory is defended against external actors such as cattle ranchers with whom conflicts over land ownership remain unresolved and latent.[5] Thus, it must be understand that the effects of disasters caused by climate change (floods and droughts) affect them directly in their subsistent economy. Two aspects should also be emphasised: (1) There is constant conflict between farmers regarding the definition of property ownership in the same space; cattle ranchers claim possession as individual property, and the indigenous as collective; (2) Cattle ranchers carry out activities such as *roza*, *tumba*, and burning which cause deforestation and climate change, thus bringing severe consequences on the indigenous population.

Ranchers and Deforestation

The rearing of livestock is concentrated in large and medium-sized properties. Farmers practicing livestock rearing are interested in the land, and the highest proportion of it. There is therefore both a commercial and maximum exploitation of the land. So the practice of *roza*-and-burn activities of cultivated areas are activities that increase with the agricultural season, in order to induce the regrowth of grasslands to feed the cattle and the elimination of weeds, or with the aim of enabling more fields of grazing. Fire is a low-cost tool, widely used for the management of grasslands and is employed in economic activities on as large a scale as subsistence.[6] In this way, fires, once started, could become uncontrollable. The indigenous and peasant people argue that the damage fires bring to them include the loss of their crops almost in its entirety, the deterioration of the land, and the loss of grazing areas, affecting the agricultural practice with livestock rearing (Baudoin 2012: 57).

Livestock remains one of the most important, direct causes of deforestation in the Brazilian and Bolivian Amazon. Ranchers, whether on medium or large scale, are the principal parties responsible for transforming forests in pastures (Jong and Mery 2011:12).

Historic Flooding in the Bolivian Amazon: Live it to Tell it

Between November 2013 and February 2014, there were very heavy rains in the basin of the Beni River. Since November, the Abuná and Madera rivers were already at levels above the historical average and by early February they had already surpassed the historical maximum, and were expected to continue to rise even more (Molina y Bustamante 2014).

As of 1 February, the whole basin was saturated and by 15 March the water level was still rising. Under such disastrous conditions, in February 2014 several municipalities in Beni called a hydrological red alert. The municipalities affected were: San Ignacio, Santa Rosa Yacuma, Trinidad, San Borja, and Rurrenabaque Reyes. In Beni alone there would have been 3,957 families affected, in addition to 140 acres of crop losses and 218 families left homeless as a result of flooding[7] (Humérez 2014; Izurieta 2014).

In the affected areas, the population is beginning experience a growing shortage of food and water, while being isolated in the Northern Amazon due to the bad state of the roads and highways. It was predictable that the prolonged flooding would generate outbreaks of disease, a change in soil composition, migration of animal species, and altering geography plus population displacement. Beni was in a state of emergency and alert which caused families to evacuate their homes. In the region of *Territorio Indígena Parque Nacional Isiboro Sécure* (TIPNIS), 100 per cent of communities in the area were flooded and residents were forced to live above two metres of water, either by barge or in the few spaces high enough to be free of the flood water. The desperate population called for greater attention to be paid to the lack of food and clean drinking water, as well as the spread of disease and the loss of almost all their cattle (24/02/14 ERBOL in Humérez 2014).

Causes of the Flood

An Already Changing Climate in Beni

Based on a study that took place between 2000 and 2011, it can be said that in Beni there has been an increase in total annual precipitation during the eleven years of study. In 2000, the total average annual rainfall was 944.8 mm, while in 2007 and 2011, a sharp increase of 168 mm was recorded which was above average. The meteorological records show that, by far, 2007 and 2009 were the wettest years, with precipitation of 1,541 and 1,814 mm / year registered, respectively (Baudoin 2012:28).

In Beni, an average of 59 days of rain/year in the 2000s was recorded. Climatic data shows a general trend of increase by 10 days of rain in the last five years. Between 2000 and 2005, it rained 48 days on average, while in 2006 and 2011, it rained 58 days. 2009 was the year it rained the most, recording a total of 72 days of rain. There was little change recorded with the number of days of thunderstorms during the eleven years of study, keeping close to the average at 37 days (Baudoin 2012:29).[8]

The results show variations in the flooded area between 2004 and 2011. On average the flooded area in the department of Beni covered about 6,000,000 hectares. The years in which the greatest amount of flooded areas during the wet season was recorded were 2004, 2006 and 2008, when approximately 12,000,000 and 9,000,000 hectares were covered by water due to overflowing rivers in the Amazon basin (Baudoin 2012: 46).

Moreover, the region is characterised by hot spots, which are related to deforestation.[9] Most fires occurred in the municipalities where the savannahs predominate.

Beni has the second largest number of hot spots in Bolivia during the dry season.[10] In July 2010, the SENAMHI[11] reports that Bolivia was the second country with the most hot spots in South America, coming only after Brazil. In August 2009, ABT[12] reported that Beni had the highest incidence of hot spots in the country, and was expected to move from orange to red alert due to an increased level of fires. The fires covered one million hectares of Bolivia. The year 2010 ended with the declaration of el Beni as an emergency zone due to drought and the fires that affected it. This declaration was maintained at least until March 2011 due to natural disasters caused by heavy rains brought by the La Niña. There were two abnormal droughts, independent of the El Niño drought, which affected the Amazon region in 2005 and 2010. Both years were particularly hot and dry. Bolivia's government declared a state of emergency in Beni in both 2005 and in 2010, the worst ranking drought since 1963 (Oxfam 2009, in Baudoin 2012: 63-65).

Owing to the high rainfall during the wet season, the region experiences annual flooding that can even extend right into the dry season (Costas and Foley 2002). In 2009, there was particularly wet and heavy flooding in the Amazon region, to the point that some authors categorised the event as one of the worst floods in the last fifty years (Chen et al 2010 Baudoin 2012: 69). And surely the floods of 2014 would be considered exceptional in the last half century.

Map 7.1: Hot spots

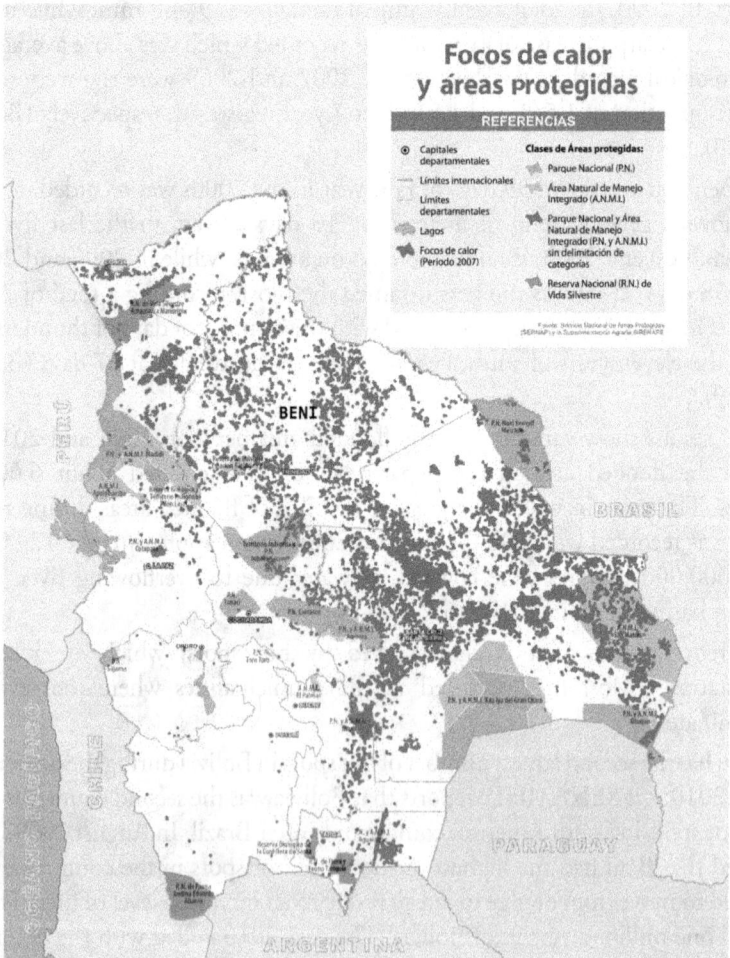

Source: CEDIB 2014

The frequency of droughts is expected to increase in the region as a result of deforestation and climate change (Williams et al 2007; Malhi et al 2002 in Baudoin 2012: 70).

Thus, we note that the flood of early 2014, an unprecedented extreme in history, has its roots in climate change and had already been manifesting itself in the first decade of the century.

Deforestation

Bolivia is the fifth country with the most deforestation in the world. Deforestation and forest degradation occur in all forest ecosystems of Bolivia, especially in the Amazon, as Urioste (2010:3) notes.

Deforestation is the main cause for the emission of greenhouse gases and, hence, the effects of climate change in Bolivia. Following this is the most frequent cause of ecosystem degradation and loss of natural capital at a rate unprecedented in Bolivian history.[13]

In Bolivia , there is a deforestation rate of 350,000 hectares per year, but in per capita terms 320m2/persona/yr, resulting in a 20 times higher rate than the global average (~16 m2/persona/yr) and is the highest in the world, surpassing the levels of other major deforesting countries. In lowlands, deforestation processes are responsible for 95 per cent of biodiversity reduction, while climate change is only responsible for 5 per cent. An expected deforestation of 33 million hectares by the end of this century means the emission of 8 billion tons of CO_2 (Urioste 2010 : 3).

The rapid increase in the frequency and intensity of fires in the Amazon has become an environmental issue, creating political and social pressure on governments to regulate the use of fire and reduce deforestation rates in the Amazon forests (Baudoin 2012: 7).

The growth process of *tumba* and burn, reflects a dramatic increase in forest clearings, parallel to major advances in agricultural frontiers for agriculture and livestock[14] (Goitia 2014: 10).

Brazilian Dams

A third cause of the floods this year falls on the two mega-dams in Brazil, Jirau and San Antonio. As Goitia (2014: 10) points out, the floods caused by water retention in dams built in Brazil, and near the border with Bolivia, will have a permanent effect, causing further loss of forests, destruction of man-made infrastructure, impacts on livestock, flora, fauna and wildlife, and impacts on human life.

Both dams began operating at their highest level, and coincided with extraordinary rains. In an area where this is uncommon, putting up a wall (construction of dams), and generating a decrease in water velocity is a monumental blunder (Archondo 2014: 4; Interview with Molina, Patricia).

In Brazil, they had already made modulations for average flow. The average flow is somewhat less than 35,000 cubic meters per second, and they were able to model up to 40,000 or 45,000, meanwhile flow this year reached 56,000 cubic meters per second or more. Their model could not measure the extent of the disaster (Archondo 2014: 5; Interview with Molina, Patricia).[15]

Note that in 2007, the Brazilian Institute of Environment (IBAMA) indicated that dams on the river Madera would have direct or indirect impacts on Bolivia (Medina 2014: 4-5).

Consequences and Conclusions

We began this chapter by noting that the effects of climate change in the Bolivian Amazon have their cruellest repercussions on the indigenous population. Amazonian families feed off birds and forest animals that are sources of protein of animal origin and complementary to nutritional needs vital for food security, and floods of 2014 have caused the decrease of species such as jochi, fishing, the *anta*, the *chancho de monte*, the *huaso*, the kettle, and so on.

Intermediate cities received the migration of rural residents affected by flooding and the outskirts of these cities are going to swell in numbers, increasing underemployment, forced labour, and begging, as the migration of the Sirionós to the city of Santa Cruz has forced people to beg.

As Cuellar (2014:9) points out, the damage to the mud houses or thin wood are irreversible. Approximately 5000 families across the region of Beni and the Northern Amazon have been left homeless.

Four fish species will go into extinction, known as ecocide, because the fish cannot spawn in the Amazon because of the presence of dams. (Interview with Teresa Flores 2014). Ramos said that some 650 fish species will disappear, affecting the fishing capacity of the region and the livelihoods of local communities, who will have to change their diet (2014: 7). Potential impacts on human health and proliferation of malaria (Medina 2014: 4-5 and Ramos 2014) were also detected.

In short, the indigenous people of the Amazon cannot face extreme situations like the floods experienced in 2014. The facts show that each time their communities are flooded, they are virtually left with nothing; they are left with an empty pantry, living each day as it comes from nature.

In the Amazon, everything should be done to reduce deforestation because by its nature, forests provide important environmental services, mitigate greenhouse effects (GHG), including carbon dioxide and others. They also maintain valuable biodiversity in the world of flora and fauna, and contribute to a decrease in global warming. These are the reasons why those parties who live off the land for commercial endeavours, should be regulated by the state with more drastic laws, otherwise the country is going to experience many more disasters such as those endured in 2014.

Several forces are shaping the Amazon region. Persistent deforestation, land degradation, poverty, and violence regarding land appropriation. Thus, these aspects mark an inequality and in the context of disasters this becomes even deeper.

Notes

1. At the national level, the media reported 59 deaths, 146 municipalities and around 60,000 families affected in addition to 110,000 heads of cattle lost.
2. Clearing land (covered by forest and vegetation) by burning the forest and vegetation cover in order to prepare the land for pastures.
3. Emergency Disasters Database
4. This is based on the ILO Convention 169 of 1989 concerning Indigenous and Tribal People, and the Declaration of the Rights of Indigenous People, adopted by the General Assembly of the United Nations in 2007. Both the ILO Convention as well as the Declaration of the United Nations establish the right of the indigenous people to self-determination and the lands, territories and resources which they have traditionally owned, occupied or otherwise used or acquired. The Declaration of the United Nations recognises the right to free, prior and informed consent on all projects affecting their territory (Feldt Heidi 2011: 5).
5. Broadly, this theme is developed in Canedo, Gabriela (2011) La Loma Santa, a gated utopia. State, Territory and culture in the Bolivian Amazon.IBIS-Plural, La Paz.
6. See José Martínez et.al 2003 Fire in the Panatanal. Forest fires and loss of biodiversity resources in San Matías-Santa Cruz, PIEB, La Paz.
7. At the national level, reports point to 59 deaths, 146 municipalities and around 60,000 affected families. In addition, and according to the latest report of 17 February by the Ministry of Rural Development and Land, 110,000 heads of cattle have been lost in the municipalities affected including a loss of approximately 39,000 hectares of crop (Humérez 2014).
8. According to Jean Luc Bourrel (1999) the dynamic of floods are considered to be of two types: the endogenous and exogenous, depending on rainfall originating from the floodplain or outside area, respectively. This feature makes the flooding process dependent on the distribution of rainfall in the upper basin (in mountains or foothills) or lower part of the basin (local rainfall in the plain). Depending on whether the rain came from the top or bottom, the hydrological response will be different. A third situation occurs when precipitation originates in both the top and the bottom, resulting in extreme flooding. On the other hand, according to Josaine Ronchail, generating floods in Beni is also dependent on other factors such as the degree of saturation of the soil: precipitation within normal values can produce extreme flooding if in the previous year there was saturation of the soil (Baudoin 2012: 7-8).
9. In reality, fire has always been associated with the dynamics of some ecosystems.For instance, throughout the world savannahs have always been associated with fire, be it of anthropogenic origin or not (Baudoin 2012: 10).
10. Points of forest fires.
11. Servicio Nacional de Meteorología e Hidrología (National Meterological and Hidrological Service).

12. Autoridad de Bosques y Tierras (Forest and Land Authority).
13. Globally, about 13 million hectares of tropical forests – that is, an area the size of Nicaragua – are lost each year to be converted to other uses. This loss represents a fifth of total global carbon emissions, which is why deforestation is considered the second most important factor of global warming. Consequently, forest conservation plays a vital role in any initiative to combat global warming.
14. In 1995 after the development of the forest map of Bolivia 53.4 million hectares was established. The deforestation was 168,000 hectares per year. In 2003 the estimated deforestation had exceeded 300,000 hectares per year and by 2010 forest loss stood at 400,000 hectares (Goitia 2014).
15. See Lanza, Gregorio; Arias, Boris 2011 Represa Cachuela Esperanza. Posibles consecuencias económicas y ambientales de su construcción. CIPCA, Cuadernos de Investigación 74, La Paz.

Bibliography

Archondo, Rafael, 2014, ¿Fueron dos represas brasileñas causa de las inundaciones en el Beni? Página Siete (11 de mayo de 2014). Interview to Patricia Molina and Abraham Cuellar.

Baudoin, Mario et al, 2012, *Inundaciones e incendios. Elementos para un acercamiento integral al problema en el Beni.* PIEB, DANIDA, La Paz.

Beck, Sthepan; Moraes, Mónica, 1997, 'Llanos de Mojos Region'. In: S. Davis et al., *Centers of Plant Diversity*, Vol 3, Oxford: The Americas.

Beck, Sthepan; Moraes, Mónica, 2004, 'Características biológicas generales de la llanura del Beni'. In: Poully, M; Beck, Sthepen; Moraes, Mónica; Ibañez, C *Diversidad biológica en la llanura de inundación del río Mamoré. Importancia ecológica de la dinámica fluvial.* Fundación Simón I, Santa Cruz: Patiño.

Canedo, Gabriela, 2011, *La Loma Santa una utopía cercada. Estado, Territorio y cultura en la amazonia boliviana,* La Paz : Ibis-Plural.

CEDIB, 2014 *Atlas de Bolivia, Tierra, territorio y Recursos Naturales,* La Paz : CEDIB.

Chevallier, Pierre; Pouyaud, Bernard; Suárez, Wilson y Condom, Thomas, 2011, 'Climate change threats to environment in the tropical Andes: glaciers and water resources'. In *Regional Enviromental Change* Volumen 11, Supplement 1.

Cuellar, Abraham, 2014, 'Impactos económicos de las Represas y alternativas económicas en el Norte Amazónico'. In *Hora 25*, No 107-108 junio de 2014. Ponencia presentada en la VI Cátedra Libre Marcelo Quiroga Santa Cruz. Paraninfo de la UMSA, lunes 31 de marzo y martes 1 de abril.

Feldt, Heidi, 2011, *Fortalecimiento de Organizaciones Indígenas en América Latina: Pueblos Indígenas y Cambio Climático. Relación entre cambio climático y pueblos indígenas y sus posiciones en el contexto de las negociaciones en la Convención Marco sobre el Cambio Climático,* Giz, BMZ, Alemania.

Francou, Bernard; Ribstein, Pierre; Wagnon, Patrick; Ramírez, Edson; Pouyaud, Bernard, 2005 'Glaciers of the Tropical Andes: Indicators of Global Climate Variability', En: *Advances in Global Change Research* Volume 23.

Gallardo, Mauricio 2012, *Pobreza y cambio climático: un análisis de equilibrio general para Honduras,* Buenos Aires : CLACSO.

García, Virginia, 2004, *La perspectiva histórica en la antropología del riesgo y el desastre. Acercamientos metodológicos.* Relaciones. Estudios de historia y sociedad, vol XXV, núm. 97, invierno, pp. 124-142, El Colegio de Michoacán, México.

Goitia Luis, 2014, 'Impacto del manejo forestal y deforestación en Bolivia'. En: Hora 25, No 107-108 junio de 2014. Ponencia presentada en la VI Cátedra Libre Marcelo Quiroga Santa Cruz. Paraninfo de la UMSA, lunes 31 de marzo y martes 1 de abril.

Humérez, Ximena, 2014, 'Lluvias no dan tregua y aumentan las cifras de damnificados en el país'. In: http://cipca.org.bo/index.php?option=com_content&view=article&id=3041:lluvias-no-dan-tregua-y-aumentan-las-cifras-de-damnificados-en-el-pais&catid=186:noticias-2014&Itemid=216 (visitada en marzo 2014)

Izurieta, Edgar, 2014, 'Beni resiste inundación y se declara alerta roja a la cuenca del Mamoré'. In: http://cipca.org.bo/index.php?option=com_content&view=article&id=3030:beni-resiste-inundacion-y-se-declara-alerta-roja-a-la-cuenca-del-mamore&catid=186:noticias-2014&Itemid=216 (visitada en marzo 2014)

Jong Wil de y Mery Gerardo, 2011, *Desafíos de los bosques amazónicos y oportunidades para el manejo forestal comunitario*, CIAS Discussion Paper 20 IUFRO Ocasional Paper 20, Finlandia.

Lanza Gregorio, Arias Boris, 2011, *Represa Cachuela Esperanza. Posibles consecuencias Económicas y ambientales de su construcción.* CIPCA, Cuadernos de Investigación 74, La Paz.

Martínez, José et al, 2003, *Fuego en el Pantanal. Incendios forestales y pérdida de recursos de biodiversidad en San Matías-Santa Cruz*, PIEB-CEDURE-Fac. de Humanidades UAGRM, La Paz.

Medina, Jorge, 2014, 'Gobierno del Brasil tomó decisión política de construir represas pese a quien pese'. In: *Hora 25*, No 107-108 junio de 2014. Ponencia presentada en la VI Cátedra Libre Marcelo Quiroga Santa Cruz. Paraninfo de la UMSA, lunes 31 de marzo y martes 1 de abril.

Molina, Patricia; Bustamante, Andrés, 2014, 'La crecida histórica del 2014 y la sincronización del desastre'. In http://www.fobomade.org.bo/art-2348 (publicado 2014-05-25)

Oviedo, Gonzalo, 2008, *Los pueblos indígenas y tradicionales y el cambio climático* (Documento de trabajo versión resumida) Unión internacional para la Conservación de la Naturaleza.

Ramos, Juan Pablo, 2014, 'La Construcción de las represas es una decisión política del Brasil'. En: *Hora 25*, No 107-108 junio de 2014. Ponencia presentada en la VI Cátedra Libre Marcelo Quiroga Santa Cruz. Paraninfo de la UMSA, lunes 31 de marzo y martes 1 de abril.

Soruco, Álvaro; Vincent, Christian; Francou, Bernard and Gonzáles, Francisco, 2008, 'Glacier decline between 1963 and 2006 in the Cordillera Real, Bolivia'. In: *Geophysical Research Letters*, Vol. 36.

UNDP, 2011, *Tras las huellas del cambio climático en Bolivia. Estado del arte del conocimiento sobre adaptación al cambio climático. Agua y seguridad alimentaria*, UNDP, La Paz.

Urioste, Andrea, 2010, *Deforestación en Bolivia* (documento de trabajo) Fundación Friedrich Eberton, La Paz.

Urioste, Miguel; Pacheco, Pablo, 2001, *Las tierras bajas en Bolivia*, PIEB, La Paz.

8

Gender-wise Rural-to-Urban Migration in Orissa, India: An Adaptation Strategy to Climate Change

Nirmala Velan & Ranjan Kumar Mohanty

Introduction

Migration caused by human action or natural hazards, or cyclical environmental factors, results in temporary or permanent dislocations of people. These displacements are caused by sudden events like flooding, earthquakes, volcanoes, hurricanes, cyclones, forest/bush fires, Tsunamis, industrial accidents or chemical leakages. These hazards affect both the livelihood and ecosystem of the area. An environmental hazard or adverse climatic change that results in immediate displacement or migration of people immediately after its occurrence is known as environmental emergency migration, as in the case of Tsunami, hurricane, flood, etc. Environmental migration is viewed as an adaptation strategy of households to either diversify or improve livelihood under constant threat of environmental change (UNDP 2009). From 2007, the IOM (2007) defines 'environmental migrants' as 'persons who, for compelling reasons of sudden or progressive changes in the environment that adversely affect their lives or living conditions, are obliged to leave their habitual homes, or choose to do so, either temporarily or permanently, and who move either within their country or abroad'. It identifies three types of environmental migrants, namely, (i) Environmental emergency migrants; (ii) Environmentally motivated migrants; and (iii) Environmentally forced migrants. However, they are more commonly called as 'environmental migrants'.

Renaud, et al. (2007) categorises environmental migrants as environmentally motivated, environmentally forced and environment refugee migrants. According to them, individuals who temporarily flee the worst environmental impact, like Tsunami, hurricane, etc., are 'environmental emergency migrants'. Whereas 'environmentally forced migration' is the compulsion to move to avoid worsening environmental deterioration. It is a relatively slower process, which may or may not leave a choice for the affected individuals to return to their original place. The third category is environmentally motivated migration, under which people move from an area of gradually deteriorating environment. The process of migration here tends to be slow. Socio-economic factors assume a significant role in this case, unlike the two previous categories. Migration processes vary in the last two categories at global level, given the difference in vulnerability of the group involved. Among the three categories, 'environmentally emergency migrants' tend to require the maximum support for suitable alternative livelihood strategies and sustainable development. Therefore, while the emergency migrants would require immediate attention, the motivated ones would need information on alternative livelihood opportunities, safety and protection, and infrastructural support.

It is quite difficult to record, isolate or pin the displacements and migration caused by environmental degradation, due to multiplicity of factors leading to migration. Policy makers are unable to frame suitable policies to effectively contain or assist environmental migrants, for want of proper definitional pinning. The need to categorise the different migration processes arises from the fact that it can help in the formulation of suitable policies at national and international levels. Policy solutions for coping, support and adaptation would differ based on the type of migration. This requires researchers to identify factors that induce and affect environment-induced migration that would contribute to the development of policy framework to establish the adaptation and migration nexus (Stal and Warner 2009). Furthermore, people who decide not to move would also require support in terms of land management techniques, besides training to adopt or change employment. This would involve consideration of the impacts on children and women especially, as environmental changes are found to influence male and female migration differently (Findley 1994; Henry, et al. 2004; and Renaud et al. 2011), which requires different policy prescription at local and international level. Additionally, the government may use legislative measures to prevent returnee migrants or original residents from staying in the affected locality due to risk of recurring hazards (as in the case of floods), by demarcating the area as a danger zone.

Although all countries and people are affected by climate change, its impact distribution tends to be unequal and skewed towards the poorest, who lack sufficient economic, technical, institutional and scientific capacity to adapt or cope. This is true of countries as well as people, of whom the poor find it hard to respond to climate change. Rural areas are the most vulnerable to climate changes, in which two-thirds of the world's poor reside with nature-based livelihood activities, and lack vital goods and services, including health (Horton et. al. 2010), education and information (Casillas and Kammen 2010). Hence, adverse climate changes like sudden flood or cyclones, impact livelihood, income and settlement, besides rural infrastructure. Heat waves and droughts generate economic stress due to reduced production and productivity. This results in unemployment leading to low and semi-skilled migration to urban areas (Gray and Mueller 2012). Thus, climate change hazards strongly influence rural poverty level. Meanwhile, it is also argued that since the rural people are often exposed to climatic risks, they tend to be more adapted to it (Nelson et al. 2010) than the urban poor (Ruel et al. 2009).

Climate change affects the basic requirements of human beings in terms of food, water, infrastructure and resources. A common trend of adaptation to climate change followed by rural people is livelihood diversification, shifting crop cultivation combinations, water harvesting or water shed, and migration, especially from rural to urban areas, all of which requires sufficient institutional support for sustainability of livelihood in the long run (Easterling et al. 2007). The International Assessment of Agricultural Knowledge, Science and Technology for Development (IAASTD 2009) stresses on research and adaptation strategies to promote participation, empowerment and social learning for rural people. The adaptation strategy is largely governed by the extent of climate change, its global inter-linkages and the resources needed for adaptation strategies. These, in turn, are likely to trigger stress tensions and even conflicts, both individual and societal, which tend to further aggravate inequity. Therefore, in such situations, coping requires socio-economic, environmental and political support. in addition to good social networks, local organisations/non-governmental organisations (NGOs) help at national or solidarity international levels in determining the strength of coping capacity (Renaud et al. 2011).

Though under-estimated in research, flooding frequently constitutes an important cause of internal displacement in the populated countries of Asia, like India, Pakistan, Bangladesh, China and Vietnam, and results in widespread severe economic, social demographic and health problems. Rising sea level and drought-induced famines also contribute to migration, largely within the country (Terminski 2012). Added to the political and economic reasons for migration in the last century, environmental and feminisation of migration at both national and international level, has captured the attention of research in recent times. In light of these issues, this paper analyses the circumstances under which an individual decides to move or not to move within the migration framework. Overall, it attempts to gauge the determinants of rural to urban migration and the adaptability of rural households under environmental change. An understanding of who migrates, under what circumstances, how far and why, would provide a deeper insight into the nature, type and cause of migration, facilitating policy making for their welfare and for those who do not migrate. Therefore, the main objectives of the study are to:

i) gain an overview the variations in socio-economic background of the respondent households by migrant status before and after migration/given period by gender in Puri district, Orissa;

ii) analyse the factors inducing gender-wise rural to urban migration among the rural households in the study area;

iii) examine the impacts of migration in terms of the benefits gained and problems experienced by the migrants and their families;

iv) survey the reasons for non-migration by gender; and

v) assess the impact of climate change on poverty and income inequality of the sample households by gender and migrant status.

The remaining paper is organised in the following manner. After the introduction in section 1, section 2 summarises some of the theoretical issues, along with a few review of related literature. Section 3 gives a brief overview of the study area and its natural calamity, followed by an outline of the data and methodology used in section 4 The empirical results are discussed in section 5, while the concluding remarks are given in section 6.

Theoretical Issues and Related Literature

Migration is one of the livelihood stress reliever strategies adopted by households, under circumstances including climate change (Barnett and Adger 2007). In recent years, climate change induced migration has emerged into a coping mechanism to deal with the risks and uncertainty among vulnerable households with low capabilities and security (Tacoli 2009). Given that climate change is going to cause worse incidents of human displacement globally (Guterres 2008a, 2008b), it is necessary to fully understand the determinants of displacement and migration caused by environmental change and degradation, and gauge the adaptability, resilience and sustainability of environmental change induced migration (Boano et al. 2008; and Barnett and Webber 2009).

In the 1970s and 1980s, researchers working on environmental hazard mainly focused on forced displacement, which drew international attention to the potential severity of the problem. Towards the late eighties, environmental displacement was mainly associated with desertification, drought and famine. It was only after the establishment of the Inter-governmental Panel on Climate Change (IPCC) in 1988, the adoption of the United Nations Framework Convention on Climate Change, and the Convention on Biological Diversity, following the Earth Summit conducted in 1992 in Rio Janeiro, followed by the United Nations Convention to Combat Desertification in those Countries Experiencing Serious Drought and/or Desertification, particularly in Africa (1994), that environmental research began to take shape in the 1990s. The subject also became the concern of several institutions, like IPCC, United Nations Environment Programme (UNEP), International Organisation for Migration (IOM), United Nations University – The Institute of Environment and Human Security (UNU-EHS 2003), and United Nations High Commissioner for Refugees (UNHCR - Terminski 2012).

The research works that followed focused on the association between climate change and migration, stressing the need for further evidence to substantiate it (Gómez 2013). The studies highlighted the difficulty of analysing the causes of environmental displacement/migration, which is influenced by various factors, comprising demographic, social, economic, political and environmental factors (Laczko and Aghazarm 2009). Initially, labour migration studies at micro- and macro-level modelling had considered only socio-economic variables influencing it, while ignoring the environmental and social influences as potential determinants. This could have been due to the methodological problems of quantifying or identifying suitable proxy variables for them (Radu 2008). The Asian Development Bank report (ADB 2009) suggested that climate change be included as one among the factors leading to migration. However, it is a known fact that it is difficult to isolate the impact of climate change, especially on the rural migration, due to the complexity and multiplicity of factors influencing it.

In addition, recent theoretical developments attempt to include the influence of social network into migration models, which contributes to the emergence of 'social multipliers or externalities' (Manski 2000; Durlauf 2001; and Glaeser and Scheinkman 2001). Its basic premise is that individual behaviour is influenced by choices made by other members of their group, giving rise to externalities resulting in population behaviour. This idea came to be incorporated into migration decision research, according to which the decision to migrate is not done by the individual in isolation, but are also determined by the actual choices of others in the group (endogenous effects) or their behaviour (contextual effects). Earlier, very little modelling was done by including these externalities, namely, migrant networks, immigrant cluster, herd behaviour, chain migration or peer influences (Radu 2008). The earliest studies that included it are by Thaddani and Taylor (1984); Chau (1997); and Helmenstein and Yegorov (2000). Recent researches, however, have theoretically and empirically demonstrated that social networks significantly influence a migrant's decision on where and when to migrate (Munshi 2003; Esptein and Gang 2004; and Bauer et al. 2006). Besides, variations in gender roles, capabilities and responsibilities are also expected to further widen disparities under climate change hazards (Vincent et al. 2010). This has been confirmed by the studies of Nelson et al. (2002), Huisman (2005) and Omolo (2011).

A few related research works on different dimensions of environmental migration may be reviewed here. Although Findley (1994) found no change in the overall migration level during the drought from 1983 to 1985 in Mali, a substantial rise in the migration of women and children was observed. A study on rural out-migration by Ezra and Kiros (2001) revealed that households affected by recurrent and severe droughts in Ethiopia adopted an income diversification strategy through migration of only some family members, with the others remaining behind.

Henry et al. (2004) analysed the association between variability in rainfall and migration in Burkina Faso, and found that while the probability of long-term female migration to other rural areas had declined during low rainfall, that of the males had increased. Meanwhile, in the case of Indian Ocean Tsunami, no long-term out-migration pattern was reported from the areas affected, although several families expressed their interest in migrating (Naik et al. 2007).

Doevenspeck (2008) examined the process and perpetuation of internal migration in rural Benin of West Africa and concluded that despite environmental problems, the affected migrants moved mainly due to socio-cultural factors, which rendered identification of the cause of migration difficult. Whereas, a review of impact of climate change by Black et al. (2008) revealed it to be aggravating prevailing problems, and increasing migration to safer locations that offer household security and reliable livelihoods. However, since migration occurred due to multiple factors, it was difficult to isolate the effects of environmental hazard from the economic factors. In developing countries, Ahmed et al. (2009) reported increase in poverty due to climate volatility, clearly evident in Indonesia, Bangladesh, Africa and Mexico.

Molaei, Santhapparaj and Malarvizhi (2008) analysed the earning gains of rural migrants settled in urban Iran and concluded that the migrants' demographic characteristics, employment sector and social network significantly influenced their earnings. Meanwhile, Ellis (2009) found that migration in general contributed to income diversification in rural areas. Groen and Polivka (2009) reported that 63 per cent of the migrated population due to Hurricane Katrina had returned to their original country after 13 months of its occurrence. The main determinants of their decision to return were age, ownership of house and severity of damage of their country. The non-returnees had the opportunity of restarting their lives, but some of them had to experience unfamiliar labour market conditions, besides the lack of support structures and social networks

In a study on Niger, Afifi (2010) observed that economic consequences of environmental change were more of a push factor to migration, calling it 'environmentally induced economic migration'. Black et al. (2011) concluded that individual migration decisions and flows were affected by economic, political, social and demographic factors operating in combination, on which the effect of the environment was highly dependent. Environmental change indirectly affected migration, and directly the hazardousness of place. Through economic drivers it affected by changing livelihoods and through political drivers by causing conflicts over resources. Kartiki (2011) examined migration in response to cyclone Aila in rural Bangladesh and found that environmental stress had affected life, shelter, livelihood, drinking water and costal defence embankment. Although climate change increased migration, it was difficult to isolate environmental pressure due to multiple pressures. Furthermore, under repeated cyclones, migration became the last survival strategy.

In effect, migration decision modelling has developed over time into incorporating environmental change, social network and gender as additional determinants of environment induced migration. However, while studies are able to substantiate income and livelihood diversification, and social network influences of migration, it has not been easy to isolate the direct effect of environmental change on it, owing to the complexity of multiple factors influencing it.

Study Area Profile and Natural Calamities

Study Area Profile

Orissa State is bounded by the Bay of Bengal on the East, with a coast line of 450 kilometres (kms). It is the ninth largest in area (4.87 %), with the eleventh largest population. According to the Planning Commission (2012), the State has a much higher (57.2 %) population living below the poverty line, than at the all-India level (37.2 per cent) during 2004-2005. Of this, 60.8 per cent poor live in rural and 37.6 per cent in urban areas. In order to achieve inclusive growth, the State has launched several employment generation and poverty alleviation programmes, like Food Security and Public Distribution System, Indira Awas Yojana (IAY), Swarnajayanti Grameen Swarozgar Yojana (SGSY), and Mahatma Gandhi National Rural Employment Guarantee Scheme (MNREGS), to generate livelihood and provide basic needs to the poor.

Spread over an area of 3,051 sq. km., Puri is one of the developed districts of Orissa, with the Bay of Bengal on its Eastern and South-Eastern part, and a coastline of approximately 151 kms. Its hottest month is May, while June to September is the South West monsoon period. Winds are quite strong, especially in the coastal regions during the months of summer and monsoon. Humidity is also high throughout the year, particularly closer to the coastal areas. The 2001 Census recorded a total population of 1,502,682 in the district, consisting of 50.80 per cent males and 49.20 per cent females, which rose to 1,697,983 in 2011 Census, with very little change in its sex composition (50.95 % and 49.05 % respectively). As much as 86 per cent of its population reside in rural areas. The combined literacy rate of Puri was 78 per cent in 2001 (88 % male and 66 % female literacy), which increased to 85.4 per cent in 2011 Census (92 % male and 79 % female literacy). The district has a low total working population

of 29.98 per cent, of which females constitute 12.6 per cent. Cultivators and
agricultural workers account for 58 per cent of the main workers, with women
workers comprising only 7.3 per cent of its total. Overall, agriculture and related
activities are dominant, making employment highly vulnerable to adverse climate
changes. Its major industrial activities are art, craft and handicrafts. Its small-scale
and cottage industries are engaged in agro-based production activities, like seafood
processing, rice milling, forest-based products and wooden furniture, leather
products, snacks, cashew processing, coir, molasses, sauce, pickles, ice cream, and
jams, jellies and squash. Owing to the coastal location of Puri, fishing industry is
also very much developed (Population Census estimates, various years).

Natural Calamities

The geo-climatic situations of Orissa are such that they induce occurrences of
multiple natural calamities like earthquake, drought, heat-wave, fire, lightning,
hailstorm, cyclone, flooding and tsunami. Recurring natural calamities have been
a major obstruction to the socio-economic development of the State. Flooding,
hailstorm, cyclone, heat-wave and fire are more frequent in the State, causing
intense misery to its people. The major rivers of the State (namely, Mahanadi,
Baitarani, Brahmani, Budhabalanga, Rushikulya, Subarnarekha, Vamsadhara)
and their tributaries, besides depressions in the Bay of Bengal, are often flooded,
making it vulnerable to devastations. The State also faces drought several times
due to the vagaries of monsoon. In addition, a large percentage of the State's area
falls under seismic vulnerability zone.

Since 2003, Orissa has faced regular recurring floods, with the worst occurring
in 2001, 2003, 2006, 2007 and 2008, during which 21 out of its 30 districts were
the worst affected. The downstream flood of September 2008 affected Cuttack,
Jajpur, Jagatsinghpur, Kendrapara, Khurda and Puri districts in the Mahanadi
basin. During June to September, 21 districts were devastated, affecting 871
villages in 145 blocks (10 in Puri district) in September. It damaged 34,437
houses (majority huts) and dislocated 772,275 people, with eight casualty.
Furthermore, 382,080.70 hectares (19 % in Puri) of Kharif crop of small and
marginal farmers were damaged, resulting in more than 50 per cent crop loss, and
14,059.31 hectares of agricultural land sand cast (2.0 % in Puri). In addition, it
caused loss of livelihood, especially the traditional crafts and handloom weavers;
damaged nets and boats of fishermen; and public infrastructure, including roads,
rural water supply, irrigation projects, river and canal embankment, drainage
system, school and other official buildings. About 30,856 people were rescued
and evacuated to safer places and kept in temporary shelters during the floods.
Food was provided through 1,118 free kitchen centres, benefiting 30,198 people
for eight days. Besides, care was taken to provide health and sanitation facilities,
safe drinking water and livestock feed. The state government is not only involved

in relief and rehabilitation works, but is also constantly focused on disaster preparedness and mitigation, so as to minimise the adverse effects of the recurring natural shocks and risks on its persistent development efforts (Government of Orissa 2009, 2010).

Data and Methodology

In 2001, Orissa accounted for 11,054,202 migrants, consisting of 22.47 per cent males and 77.53 per cent females, indicating feminisation of migration in the State. Intra-State migration dominates total migration (93.58 %), with inter-State migration comprising only around 6.0 per cent of it (Population Census estimates 2001). The present study was conducted in the 2008 flood-affected Rupdeipur Gram Panchayat of Pipli block in Puri district, Orissa, from June-July 2009. A random sample of four villages was selected for the study, namely, Alapur, Nahamanga Patana, Kolitara and Panda Sahi. These villages experience adverse climatic change effects, like cyclonic flooding and droughts, and record a heavy incidence of migration, which is often environmentally induced. It is worth noting that some members of families in the sample villages have been migrating seasonally or temporarily for diversification of livelihood, even without the climatic changes aggravating their socio-economic conditions, due to poverty, landlessness, smaller land holdings or seasonality of agricultural employment. The presence of a known person, friend or relative in another place makes it much easier for them to migrate, resulting in chain migration.

The data have been directly collected by the researchers through personal interview method using pre-tested schedule from a random stratified sample of an experimental group of 120 migrants and a control group of 120 non-migrants having similar developmental background from the sample villages. This contributed to a total sample of 240 respondents, comprising 60 males and females each under the two categories. Respondents from both migrant and non-migrant households have been selected in order to make a comparison of their socio-economic background, that contributes to their decision to migrate or not. Further, while changes in living conditions of the migrants and their families left behind in villages have been examined across the pre- and post-migration periods; information for the non-migrants the survey period has been compared with their living status three years prior to the interview. Therefore, it was ensured that the migrant respondents selected should also have migrated for a minimum of three years, a period sufficient enough for the socio-economic impacts of migration to become evident. Further, in the study, a migrant household by gender is identified as the one consisting of at least one or more migrant male or female members at the time of interview. Contrarily, in the case of the non-migrants, the household is classified as male and female on the basis of their respective dominance in earnings and household decision-making.

The objectives for the study have been examined using simple averages, ratios, percentages, logit multiple regression, t-test, Garret ranking technique (Garret and Woodworth 1969), Standard of Living Index (SLI – Roy, Jayachandran and Banerjee 1999), Lorenz curve and Gini index. The multiple regressions have been estimated within the migration decision framework for male, female and combined samples separately, to examine the varying factors influencing their migration decision. Demographic, socio-economic, environmental and social capital variables have been included as the explanatory variables in the model. The estimated logit multiple regression equation is as given:-

$$MGRNi = b0 + b1AGER + b2EDCN + b3MART + b4DEPR + b5LAND + b6ALDB + b7HHYB + b8SLIB + b9HDTB + b10SCLK + b11CLMT + b12GNDR + U$$

where,

MGRNi = gender-wise migration dummy, taking value one if migrant and zero otherwise (i = male, female and combined respondents);

AGER = age of the respondent in years;

EDCN = education of the respondent in years;

MART = marital status dummy, taking value one if married and zero otherwise;

DEPR = dependency ratio (computed as number of non-working members divided by number of working members);

LAND = value of land owned by the respondent household in rupees;

ALDB = annual labour days worked by the respondent before migration/three years;

HHYB = household income per month in rupees before migration/three years;

SLIB = standard of living index of the respondent household before migration/ three years;

HDTB = household debt in Rupees before migration/three years;

SCLK = social capital dummy, taking value one for presence of known friends or relatives at destination and zero otherwise;

CLMT = degree of adverse effect of climatic change experienced;

GNDR = gender dummy, taking value one for male and zero for female; and

U = error term.

Step-wise procedure has been adopted in the estimation of the equation, in order to overcome the chances of emergence of multicollinearity problem, if any. The theoretically expected relationship between the dependent and independent variables are as outlined.

Migration tends to decline with an increase in age, owing to the desire to settle down peacefully as one gets older. Hence, the expected association between age of the respondent (AGER) and the dependent variable is negative.

A rise in education level (EDCN) is likely to increase migration if job prospects are good at target destination; otherwise, its expected impact is negative.

Marital status of the respondent (MART) is expected to encourage migration due to increased family responsibilities, especially for the males. Whereas for females, it is likely to have a reverse effect due to different roles assumed by them as a home-maker, care taker and child-bearer. Likewise, higher dependency ratio (DEPR) in the family is also hypothesised to have a similar impact on migration for the same reasons.

On the other hand, higher land value (LAND), representing larger land ownership by the respondent household, is expected to discourage migration due to the availability of sufficient earnings from and work in own land.

Increase in annual number of days worked (ALDB) in own village before migration/three years is hypothesised to reduce migration, due to the availability of sufficient employment opportunity in native place. Similarly, higher household monthly income (HHYB) and standard of living before migration/three years (SLIB) are expected to discourage migration, due to the economic soundness of the family. Standard of living index score (SLI) is a measure of material possessions of a household, reflecting its economic well-being. Weights are assigned to each good and amenity possessed by the household, which are summed up to obtain the total score. These scores are then classified into three SLI categories, viz., low (0-9), medium (10-19) and high (20 and above) standard of living households (for computation of SLI, refer Roy, Jayachandran and Banerjee 1999).

Large household debt in Rupees (HDTB) before migration/three years is hypothesised to increase migration, due to the economic pressure to repay loan.

Presence of known friends or relatives in other places, reflecting social capital (SCLK), is expected to encourage migration by generating chain migration effect.

The paper makes an attempt to isolate the perceived effect of degrees of environmental hazard suffered by the sample households, by including climate change as a determinant of migration. The variable is expressed as a scale, with the lowest degree of impact taking value 1, reserving 2 for relatively severe effect, and 3 for very severe effect. Rise in the degree of adverse climatic change (CLMT) impact experienced by a household is expected to encourage migration as both livelihood and survival strategies.

Being a male (GNDR) is hypothesised to increase migration due to the socio-cultural gender norms more commonly prevalent in rural areas that encourage male but restrict female mobility.

Garrett Ranking Technique (Garret and Woodworth 1969) has been applied to rank on a priority basis the push and pull factors, and problems of migration, besides the reasons for non-migration. The percentage position of each item is computed using the following formula:

$$\text{Percentage position} = \frac{100*(Rij - 0.5)}{Nj}$$

where,

Rij = rank allotted to the ith factor by the jth individual; and

Nj = total number of factors ranked by the jth individual.

The percentage position thus arrived at is converted into scores by using Garret's table. These scores of all respondents for each factor are then added up and divided by the total number of respondents who had responded, to obtain the mean score for each item. These mean scores are again arranged in a descending order and ranks allotted.

Further, besides Gini index, Lorenz curve has been used to illustrate the income inequality across gender-wise migrant and non-migrant households.

Results and Discussion

Table 1 presents the socio-economic background of the sample respondents by migrant status and gender during the survey. The mean age of the migrant male and female respondents is lower than that of the non-migrants, indicating that younger people have a greater tendency to migrate. In general, education level is observed to be very low among the sample respondents, with its average for the migrants (primary school) much lower compared to that of the non-migrants (middle school). This could be because there are no High Schools in the sample villages, besides the lack of importance attached to education.

More than half the sample respondents are married, with their percentage being much higher among the non-migrants (58 % males and nearly 77 % females). Furthermore, more non-migrants live in nuclear families (58 % males and 45 % females), than the migrants. Their percentage is the lowest among the female migrants (27 %), who prefer living in joint families, so that their family members have the support of their parents or in-laws. The average family size is around six and dependency ratio around two, regardless of gender and migration status, though their numbers are slightly more in the migrant households.

As regards land ownership, relatively more non-migrants than the migrants and more males than the females own land. About 75 per cent of the non-migrant males, as against 63 per cent migrant males own land. Whereas among the females, 57 per cent non-migrants against only 15 per cent migrant households own land. The average land value of the non-migrant males is as high as Rs. 804,050, whereas that of their migrant counterpart is only worth Rs. 93,667. While land values of the female households are also much lower compared to that of males', the land value of migrants (Rs. 5,800) is relatively much lower than for the non-migrants (Rs. 56,000). Thus, low land holdings could also have been one of the main drivers of the decision to migrate in the sample villages.

Table 8.1: Descriptive Statistics

Sl. No.	Variables	Male		Female	
		Migrant	Non-Migrant	Migrant	Non-Migrant
1.	Age in years	32.30 (7.62)	33.67 (8.44)	31.48 (8.99)	36.32 (10.11)
2.	Education in years	2.50 (0.58)	7.78 (2.08)	4.47 (1.94)	6.38 (3.40)
3.	Marital status (married = 1; others = 0)	51.66 (0.38)	58.33 (0.49)	55.16 (0.51)	76.77 (0.63)
4.	Family size (No.)	6.08 (0.77)	5.95 (0.75)	5.71 (0.75)	5.55 (1.06)
5.	Type of family (nuclear = 1; joint = 0)	0.44 (0.50)	0.58 (0.53)	0.27 (0.26)	0.45 (0.50)
6.	Dependency ratio	2.23 (1.15)	2.45 (1.10)	1.48 (0.54)	2.23 (1.30)
7.	Land ownership (yes = 1; no = 0)	0.63 (0.49)	0.75 (0.44)	0.15 (0.36)	0.57 (0.50)
8.	Land value in Rupees	93667 (38483)	804050 (80037)	5800 (17175.86)	56000 (78703)
9.	Total family income per month	3020.00 (1012)	3778.00 (1500)	2097.00 (578.41)	4001.70 (2720.76)
10.	Household debt in Rs. (before)	48633 (59832.53)	28833 (57234.03)	34583 (19956.98)	25017 (13678.71)
11.	SLI (before)	8.47 (2.66)	10.54 (2.64)	9.13 (2.50)	10.28 (3.99)
12.	Remittance per month in Rs.	3452.00 (1025)	–	847.50 (246.06)	–
13.	Per cent of income remitted	65.32 (4.99)	–	41.77 (9.56)	–
14.	Social capital - relatives/ friends at destination (yes = 1; no = 0)	0.87 (0.34)	0.34 (0.12)	0.78 (0.42)	0.16 (0.07)

Note: Brackets show standard deviation; 1 US $ = Rupees 48 during the survey.

The average monthly household income of migrant males (Rs. 3020) and females (Rs. 2097) was comparatively lower in the pre-migration period, than those of their non-migrant counterparts (Rs. 3778 and Rs. 4001.70 respectively) three years back. During the same period, the migrant households had belonged to low SLI category, whereas the non-migrant households belonged to the lower medium category. Further, the migrants and male households in general had relatively higher debts before migration/three years than the non-migrants and female households, which are again the potential push factor.

The average remittance per month by male migrants to their family is Rs 3452 and by the females Rs. 847.50, accounting for almost two-thirds and 41.77 per cent of the earnings remitted to their respective families in native villages.

As regards social capital, which significantly contributes to chain migration, 87 per cent males and 78 per cent females claimed to know a friend or some known person at the urban destination. In the case of non-migrants, their respective percentages are as low as 34 and 16.

In effect, migrants are relatively socially and economically worse-off than the non-migrants, which is one of the reasons for their decision to migrate. And by gender, female households are more affected than those of the males.

Table 2 illustrates the gender-wise occupational distribution of sample migrants before and after migration, and for the non-migrants before and after a minimum period of three years. The sample migrant and non-migrant males were largely temporarily (56.67 % and 55 %) and seasonally (43.33 % and 41.67 %) employed, respectively, before migration/three years prior to the survey. But after migration, 86.67 per cent of the former are temporarily and the rest permanently employed. Whereas, among the non-migrants not much change was observed at the time of interview, of whom 50 per cent males have temporary, 41.67 per cent seasonal and the rest permanent employment. In the case of females, only 16.67 per cent migrants were seasonally employed before migrating, but post-migration, 93.33 per cent have temporary employment, while the rest are seasonally employed. Of the non-migrants, only 8.33 per cent were seasonally employed before three years, which rose to nearly 11.67 per cent after it.

Occupational structure of the male migrants in the pre-migration period shows that 35 per cent were agricultural labourers, 20 per cent construction and related workers, and nearly 17 per cent each farmers and unemployed, while the rest were businessmen and service sector employees. Post-migration, all males are employed as construction and related workers (100 %). In the case of females, only five per cent were agricultural labourers and nearly two per cent engaged in construction and related works before migration, with as much as 70 per cent being housewives and more than 23 per cent unemployed. But post-migration, nearly 97 per cent are employed as construction and related workers, and the rest in service the sector.

As regards the non-migrants, around one-third of the males (33.34 %) were unemployed three years prior to the survey, about 28.33 per cent each were farmers and agricultural labourers, and five per cent each were construction and related workers, and businessmen. But at the time of interview, 35 per cent males are construction and related workers in their native village. Of the rest,

Table 8.2: Gender-Wise Occupational Structure by Migrant Status

Sl. No.	Details	Male Before	Male After	Female Before	Female After	Total Before	Total After
A.	Non-Migrants						
1.	Type of employment:						
	a) Seasonal	33(55.00)	25 (41.67)	5 (8.33)	7 (11.67)	38 (31.67)	32 (26.66)
	b) Temporary	25(41.67)	30 (50.00)	-	-	25 (20.83)	30 (25.00)
	c) Permanent	2(3.33)	5 (8.33)	-	-	2 (1.66)	5 (4.16)
	Total	60 (100)	60 (100)	5 (8.33)	7 (11.67)	65 (54.16)	67 (55.82)
2.	Occupation:						
	a) Farmer	17 (28.33)	17 (28.33)	-	-	17 (14.16)	17 (14.16)
	b) Agricultural labour	17 (28.33)	8 (13.33)	3 (5.00)	2 (3.33)	20 (16.67)	10 (8.33)
	c) Construction and related work	3 (5.00)	21 (35.00)	-	5 (8.33)	3 (2.50)	26 (21.67)
	d) Business	3 (5.00)	6 (10.00)	-	-	3 (2.50)	6 (5.00)
	e) Service	-	4 (6.67)	-	4 (6.67)	-	8 (6.67)
	f) Housewife	-	-	44 (73.33)	42 (70.00)	44 (36.67)	42 (35.00)
	g) Unemployed	20 (33.34)	4 (6.67)	13 (21.67)	7 (11.67)	33 (27.50)	11 (9.17)
	Total	60 (100)	60 (100)	60 (100)	60 (100)	120 (100)	120 (100)

B.	Migrants						
1.	Type of employment:						
	a) Seasonal	26 (43.33)	-	10 (16.67)	4 (6.67)	6 (30.00)	4 (6.67)
	b) Temporary	34 (56.67)	52 (86.67)	-	56 (93.33)	34 (28.33)	108 (90.00)
	c) Permanent	-	8 (13.33)	-	-	-	8 (13.33)
	Total	60 (100)	60 (100)	10 (16.67)	60 (100)	70 (58.33)	120 (100)
2.	Occupation:						
	a) Farmer	1 (16.67)	-	-	-	10 (8.33)	-
	b) Agricultural labour	21 (35.00)	-	3 (5.00)	-	24 (20.00)	-
	c) Construction and related work	12 (20.00)	60 (100)	1 (1.67)	58 (96.67)	13 (10.83)	118 (98.33)
	d) Business	5 (8.33)	-	-	-	5 (4.17)	-
	e) Service	2 (3.33)	-	-	2 (3.33)	2 (1.67)	2 (1.67)
	f) Housewife	-	-	42 (70.00)	-	-	-
	g) Unemployed	10 (16.67)	-	14 (23.33)	-	24 (20.00)	-
	Total	60 (100)	60 (100)	60 (100)	60 (100)	120 (100)	120 (100)

Note: Brackets show column percentages.

28.33 per cent are farmers, 13.33 per cent agricultural labourers, and 10 per cent businessmen. The remaining 6.67 per cent each are in service sector or still unemployed. As for their female counterparts, only five per cent were engaged as agricultural labourers, while the rest were housewives (73 %) and unemployed (21.67 %) before. Not much had changed for them in three years of the interview. Only 8.33 per cent are engaged in construction and related works, 6.67 per cent in service sector and merely 3.33 per cent as agricultural labourers. The remaining 70 per cent are still housewives and 11.67 per cent unemployed. Thus, not much change is visible in the occupational structure of the non-migrants by gender over the three year period in their native village.

Table 3 shows the perceived degrees of adverse effects of climate changes suffered by the sample households by gender and category during the June-September 2008 and earlier floods. As Puri is located in a coastal area, all households reported having suffered varying degrees of adverse effect of the climate change depending on the proximity of their village to the coast. More than 70 per cent of all sample households experienced very severe impact of the adverse climatic changes, excepting non-migrant female households of whom 53.33 per cent had suffered. The impacts were less severe for less than 10 per cent and relatively severe in all cases, wherein the percentage of non-migrant females was comparatively more. Majority of them had lost their houses, livestock and livelihood in the floods. The government had provided them with food and drinking water, and alternate shelter until the problem receded.

Table 8.3: Perceived Degree of Climatic Change Effects

Sl. No.	Degree of Adverse Effect	Male	Female
A.	Migrant Household		
1.	Less severe	9 (15.00)	5 (8.33)
2.	Relatively severe	9 (15.00)	8 (13.33)
3.	Very severe	42 (70.00)	47 (78.34)
	Total	60 (100.00)	60 (100.00)
B.	Non-Migrant Household		
1.	Less severe	8 (13.33)	10 (16.67)
2.	Relatively severe	7 (11.67)	18 (30.00)
3.	Very severe	45 (75.00)	32 (53.33)
	Total	60 (100.00)	60 (100.00)

Note: Brackets show column percentages.

Gender-wise details of migration pattern of the sample migrants is furnished in Table 4. The destination of migration is urban for all the sample migrants by gender. Majority of the males (91.67 %) and females (75 %) have migrated to the neighbouring state of Andhra Pradesh to work in Vijaywada Thermal Power Station (VTPS), located at a distance of 600-900 kms from their villages. The nature of migration reflects chain migration, with most of them following their friends or other known persons there. The rest have migrated to other urban areas within the state, travelling a distance of less than 300 kms (females 25 %) and 300-600 kms (males 8.33 %). A few males also migrated to a longer distance of 900-1200 kms.

Table 8.4: Gender-wise Details of Migration

Sl. No.	Details	Male	Female	Total
A.	Nature of Migration			
1.	Rural to Rural	-	-	-
2.	Rural to Urban	60 (100.00)	60 (100.00)	120 (100.00)
B.	Destination of Migration			
1.	Inter-state	55 (91.67)	45 (75.00)	100 (83.33)
2.	Intra-state	5 (8.33)	15 (25.00)	20 (16.67)
	Total	60 (100.00)	60 (100.00)	120 (100.00)
C.	Distance Migrated (km)			
1.	Below 300	-	15 (25.00)	15 (12.50)
2.	300-600	5 (8.33)	-	5 (4.17)
3.	600-900	52 (86.67)	45 (75.00)	97 (80.83)
4.	900-1200	3 (5.00)	-	3 (2.50)
	Total	60 (100.00)	60 (100.00)	120 (100.00)

Note: Brackets show column percentages.

Table 5 presents the logit regression results of the estimated migration decision function by gender. It reveals an increase in age of the respondent to be negatively affecting the decision to migrate across gender. This is because as individuals grow older, their inclination to migrate declines. However, the influence of the variable emerges insignificant in all cases.

Regardless of gender, a rise in the respondent's education level is found to significantly reduce the probability of migration. Given that the mean education years are quite low among the sample respondents, any further improvement in their education would discourage them from migrating to take up manual jobs.

Table 8.5: Logit Regression Results – Migration Decision Function

Sl. No.	Variables	Male	Female		Combined	
		Model-I	Model-I	Model-II	Model-I	Model-II
1.	Constant	-100.425 (10.80)*	-2.894 (0.47)	-4.549 (1.27)	-17.138 (16.77)*	-18.551 (21.07)*
2.	AGER	-0.042 (0.12)	-0.040 (0.84)	-0.026 (0.39)	-0.008 (0.09)	-0.010 (0.15)
	EDCN	-1.546 (9.51)*	-0.322 (5.82)*	-0.314 (6.65)*	-0.315 (10.94)*	-0.341 (13.87)*
4.	MART	-1.402 (0.50)	0.438 (0.35)	-0.750 (1.08)	0.379 (0.76)	0.552 (1.67)***
5.	DEPR	3.879 (3.99)*	-0.174 (0.07)	-0.026 (0.002)	0.094 (0.06)	0.176 (0.23)
6.	LAND	-0.265 (4.46)*	-0.230 (5.70)*	-0.225 (6.89)*	-0.095 (5.49)*	-0.101 (6.63)*
7.	ALDB	-0.012 (0.85)	-0.010 (10.50)*	-0.012 (16.27)*	-0.007 (12.19)*	-0.008 (15.86)*
8.	HHYB	14.237 (13.89)*	-0.436 (2.30)**	-0.451 (2.76)*	2.783 (30.16)*	2.994 (37.79)*
9.	SLIB	-1.289 (12.30)*	0.341 (5.69)*	0.292 (4.88)*	-0.191 (6.24)*	-0.210 (8.17)*
10.	HDTB	-0.199 (2.28)**	-0.007 (0.01)	-0.045 (0.29)	-0.021 (0.20)	-0.039 (0.72)
11.	SCLK	3.210 (4.53)*	2.294 (6.46)*	-	1.567 (12.75)*	-
12.	CLMT	-0.104 (0.02)	0.401 (0.45)	1.374 (9.07)*	0.095 (0.12)	0.406 (2.49)**
13.	GNDR	-	-	-	-0.370 (0.50)	0.015 (0.001)
	Log-likelihood	37.33	68.80	76.71	188.55	202.28
	Pseudo R2	0.88	0.72	0.70	0.60	0.56

Note: Brackets show t–value; and*, ** and *** indicate significance at 1, 5 and 10% levels respectively.

Contrary to expectation, marital status is observed to be negatively associated with the male decision to migrate, while it emerges indeterminate for the females. However, in both cases, the effect is insignificant. Whereas under combined analysis, the relationship is positive and significant under model-II, which implies that being married increases family economic responsibility, and hence encourages migration.

Increased number of non-working dependents in the family significantly encourages male migration, owing to higher economic burden. On the other hand, for females, the association is negative and insignificant. This is attributable to the differential gender roles prevailing in the social system, which expects them to be care-takers if the dependents are children, the sick or elderly. However its insignificance is reflective of the fact that these norms may weaken under economic pressure or over time. Meanwhile in the case of combined sample, the relationship emerges positive but again insignificant.

Regardless of gender, land ownership is observed to significantly discourage migration, implying that individuals with own land might not be willing to migrate in search of alternative livelihood elsewhere due to sufficient employment and earnings in their own land. Likewise, increased pre-migration/three years period annual days of work in the native village also negatively and significantly affect the decision to migrate across gender. This indicates that if the respondents are gainfully employed for sufficient number of days in their own village, they may not be willing to move out.

Higher monthly household income before migration or over the past three years significantly encourages the male and combined sample respondents to migrate. This could be because the mean income of the sample respondents was quite low then, and therefore more paying jobs through migration is favoured. Whereas in the case of females, it significantly discourages migration, implying that they would prefer not to migrate if the household income level is already high enough in the home village.

Higher pre-migration period standard of living is found to significantly discourage migration among the male and combined respondents, who might not wish to move away from family if the SLI is already better. However in the case of females, the relationship emerges positive and significant, implying that they would still migrate to further improve their household living standards.

Contrary to the hypothesised association, the relationship between pre-migration household debt and migration decision emerges negative across gender. But, the association is significant only for the males.

Increased social capital is found to strongly influence migration decision by gender, indicating that the presence of a friend or known person at the destination significantly induces migration, as it makes it easier to know about the work place condition, employer, nature of work, wages, accommodation and other facilities.

Adverse climatic changes are found to encourage female and combined migration, which emerges significant in model-II, when social capital variable is dropped from the regression under step-wise analysis. This clearly implies that the sample respondents prefer to move to safer locations under environmental threat, both for survival and livelihood. On the other hand, the association emerges negative, but insignificant for males, which could be because the males tend to migrate even otherwise.

The inclusion of gender into the model indicates an indeterminate and insignificant influence of being a male on the decision to migrate.

The pseudo R2 values indicate that the included independent variables in the regression models explain 56 to 88 per cent of the variations in the dependent variable.

Table 6 records the ranked push and pull factors of migration by gender.

Table 8.6: Gender-wise Push/Pull Factors of Migration

Sl. No.	Details	Male			Female		
		Total Score	Mean Score	Rank	Total Score	Mean Score	Rank
A.	Push Factors						
1.	Lack of infrastructural facilities	1623	28.47	8	1521	35.37	5
2.	Lack of non-agriculture and regular employment	4081	68.01	2	1438	28.76	8
3.	Low income in village	3713	61.88	3	1215	63.94	2
4.	Presence of known person	1410	39.16	6	422	32.46	6
5.	Poverty	4325	72.08	1	4768	79.46	1
6.	Friends moved	2491	45.29	4	1527	30.54	7
7.	To repay debt	1961	40.85	5	2912	49.35	4
8.	Need to save for future	1816	30.26	7	3204	53.4	3
B.	Pull Factors						
1.	Better income	4700	78.33	1	4596	76.6	1
2.	Better medical facilities	2337	38.95	4	2914	48.56	3
3.	Food and accommodation at work place	2827	47.11	3	2221	37.01	4
4.	Better job opportunity	3870	64.50	2	3608	60.13	2
5.	Better infrastructure	1744	38.48	5	1441	30.65	5

The foremost push factor for both the male and female respondents is poverty, caused by various factors like no or small land holding, and adverse climate changes. For males, the second push factor is lack of non-agricultural and regular employment, followed by low income in native village. The priorities appear to be different for females, who ranked low income second and the need to save for future third. Existence of social capital, in terms of friends moved and presence of known person at destination of migration, are ranked fourth and sixth by males, while the females ranked them seventh and sixth, respectively. The next more important push factors for the females are repayment of household debt (fourth) and lack of infrastructure in home village (fifth). On the other hand, for males, repayment of debts ranked only fifth; the need to save for the future seventh; and lack of infrastructure eighth. Meanwhile for the females, lack of non-agricultural and regular employment ranked only eighth.

As regards the pull factors, better income, followed by better job opportunities are the main pull factors for both the sample male and female migrants. Food and accommodation at the work place is ranked third by the males, while the females ranked it fourth. For the latter, better medical facilities ranked third, while the males ranked it fourth. Finally, better infrastructure has been ranked fifth by both the sample males and females.

The ranked reasons for non-migration by gender are given in Table 7.

Table 8.7: Gender-wise Reason for Non-Migration

Sl. No.	Details	Male			Female		
		Total Score	Mean Score	Rank	Total Score	Mean Score	Rank
1.	Sufficient-employment opportunities in village	2637	47.94	3	434	72.33	2
2.	Own landholding, assets and agriculture cultivation	4018	66.96	1	1751	33.67	5
3.	Kinship around	1510	35.95	4	2900	48.33	4
4.	Marriage	342	22.36	6	2916	72.99	1
5.	Education and health facilities available	422	32.46	5	135	27.00	6
6.	Sentimental attachment	3195	57.65	2	2813	49.35	3

The males ranked own landholding, assets and agricultural cultivation as the first reason for non-migration. However, this appears less important for the females, who assigned it the fifth rank. This could be because, in Indian society, women are seldom property owners of any kind. Rather for them, marriage comprises the foremost reason for non-migration, which holds the last priority (sixth) for the males. This is attributed to the fact that females are more strongly culture-bound in rural areas, for whom the husband's home becomes the ultimate place of stay after marriage, unless the spouse/family decide to move.

Sentimental attachment to the village is ranked second by the males, while the females ranked it third. Availability of sufficient employment opportunities in the village is ranked third by the males, which is ranked second by the females. Kinship around is ranked fourth by both the males and females. The males ranked the availability of education and medical facilities in the village fifth, which is ranked last by the females. Thus, wide variations are evident in the priorities attached to non-migration by gender, largely governed by gender socialisation in the State.

Table 8 shows the gender-wise changes in work and living conditions of the sample respondents across the pre- and post-migration periods. The data for non-migrants are not reported in the table, as very little change was observed in their conditions across the three-year period considered. The average number of hours worked per day by males before migration was 5.50 hours, which significantly rose to around eight hours post-migration. Likewise for females, the work hours rose significantly from a mere 2.12 hours to about eight hours per day. The average number of days worked per month significantly increased from around 17 to 26 days for the former, and from merely seven days to almost 25 days for the latter. Their respective average gainful employment increased significantly from almost six and only two months in their villages before migration to nearly 10 months each per annum post-migration.

The migrant's average monthly income, which was Rs. 1520 for males and only Rs. 356 for females, significantly rose to Rs. 5310 and Rs. 2038, respectively. Correspondingly, their monthly household income significantly increased from an average of Rs. 2890 and Rs. 2160 to Rs. 7460 and Rs. 4574, respectively.

The average monthly expenditure of male households rose significantly from Rs. 2890 to Rs. 5210, and for females from Rs. 2160 to Rs. 3800 post-migration. As regards personal monthly savings, while it almost doubled for the males (Rs. 647) from Rs. 306 during the pre-migration period, it increased from a mere Rs. 8.33 to Rs. 233 post-migration for the females. This added to an increase in their respective monthly household savings from Rs. 492 to Rs. 877, and from Rs. 8.33 to Rs. 330.

The pre-migration period debt amount was substantially higher than in the post-migration period, and comparatively more for the male than female households. While the household debt was as high as Rs. 48,633 for males and Rs. 34,583 for females in the pre-migration period, it significantly declined to Rs. 28,289 and Rs. 14,083 respectively during the post-migration period. Thus, the migrants contributed significantly to repaying their household debts.

Table 8.8: Comparative Pre and Post-Migration Work and Living Conditions

Sl. No.	Variables	Male			Female		
		Before	After	t-value	Before	After	t-value
1.	Hours of work/day	5.40	8.53	9.38*	2.12	7.78	12.48*
2.	No. of days worked per month	17.33	25.80	8.13*	7.12	24.78	13.95*
3.	No. of months worked/year	5.60	9.83	11.94*	2.20	9.77	16.59*
4.	Migrant monthly income (Rs.)	1520	5310	27.27*	356	2038	46.78*
5.	Monthly household income (Rs.)	3020	7460	23.02*	2097	4574	36.47*
6.	Monthly household expenditure (Rs.)	2890	5210	20.81*	2160	3800	34.96*
7.	Debt amount of household (Rs.)	48633	28289	-7.41*	34583	14083	-13.06*
8.	Migrant monthly savings (Rs.)	306	647	3.76*	8.33	233	11.26*
9.	Household monthly saving (Rs.)	492	877	11.39*	8.33	330	17.53*
10.	Standard of Living Index	8.47	30.93	51.08*	9.13	22.85	28.78*

Note: * and ** indicate significance at 1% and 5% respectively.

'urther, the SLI indicates that both the male and female households that belonged o low standard of living category, significantly moved into high standard of iving category, which rose much faster for the former than for the latter in the 'ost-migration period.

The ranked problems of having migrated by gender are shown in Table 9.

Table 8.9: Problems of Having Migrated by Gender

Sl. No.	Details	Male			Female		
		Total Score	Mean Score	Rank	Total Score	Mean Score	Rank
1.	Family left behind	4708	78.46	1	4607	76.78	1
2.	Double expenditure	3704	61.73	3	4067	67.78	3
3.	No known person	1510	35.95	9	698	53.69	6
4.	Away from own community people	2575	42.91	8	2825	1	7
5.	Language problem	3374	56.23	4	3208	69.73	2
6.	Wage not received in time	3143	52.38	5	1342	22.36	10
7.	Accommodation problem	2637	47.94	6	1581	26.79	9
8.	Food problem	2168	44.24	7	3266	54.43	5
9.	Loneliness	4611	76.85	2	3580	62.80	4
10.	Risky job	695	33.09	10	1159	34.08	8

The major problem quoted by the male and female migrants is family left behind. The second problem for males is loneliness, which is ranked fourth by the females. For both, double expenditure for maintenance comprises the third main problem. Language problem is ranked fourth by the males, whereas for females it is the second major problem.

Wage not received in time is ranked fifth by the males, while the females ranked it tenth. Accommodation and food problems are ranked sixth and seventh by the former, and fifth and ninth by the latter, respectively. Staying away from own community people is ranked eighth by the males, and seventh by the females. No known person and risky job are assigned ninth and tenth ranks by the males, whereas the females ranked them sixth and eighth, respectively. Thus, male and female migrants experience different problems, governed by their respective socialisation.

Figure 8.1: Male Migrants and Non-migrants

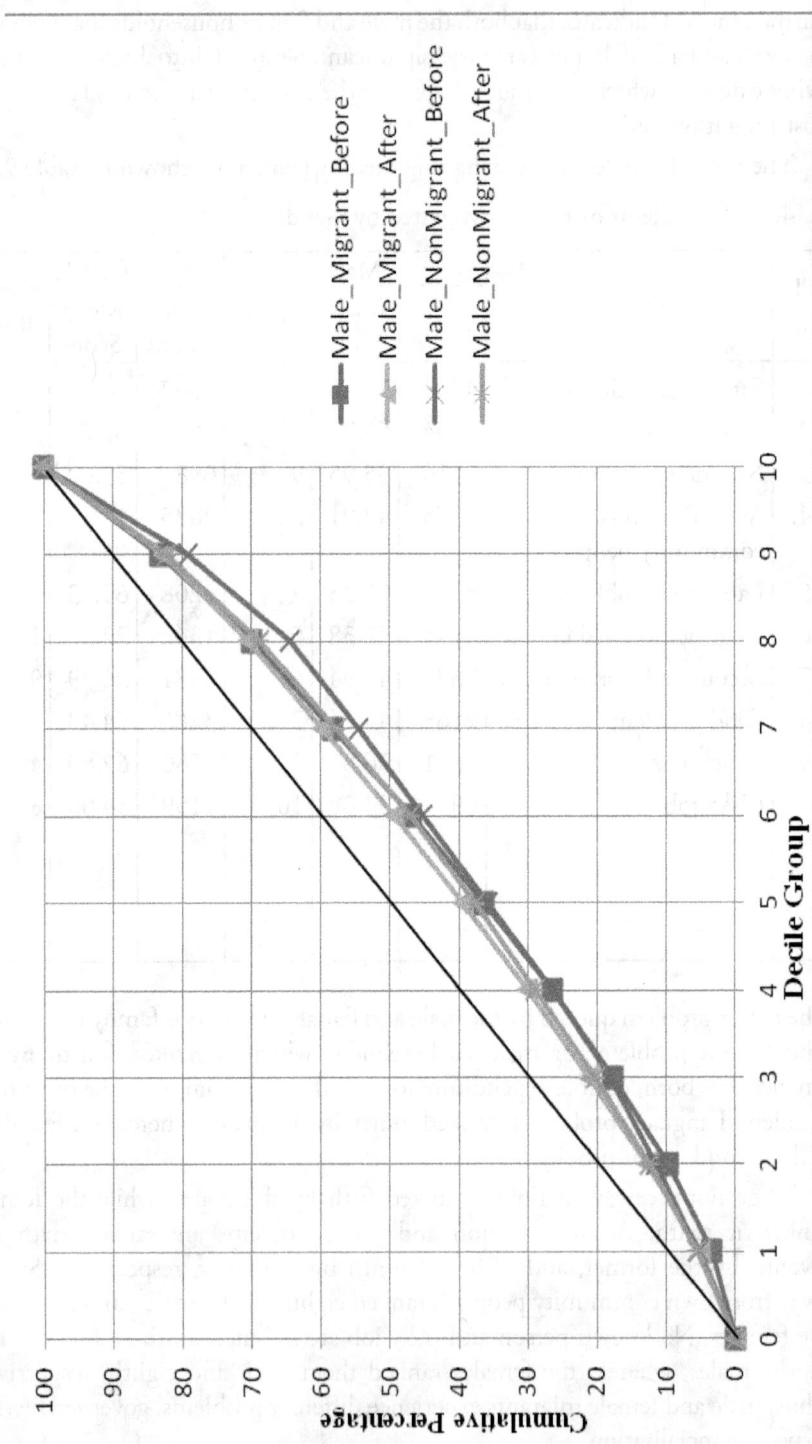

Legend:
- Male_Migrant_Before
- Male_Migrant_After
- Male_NonMigrant_Before
- Male_NonMigrant_After

X-axis: Decile Group (0–10)
Y-axis: Cumulative Percentage (0–100)

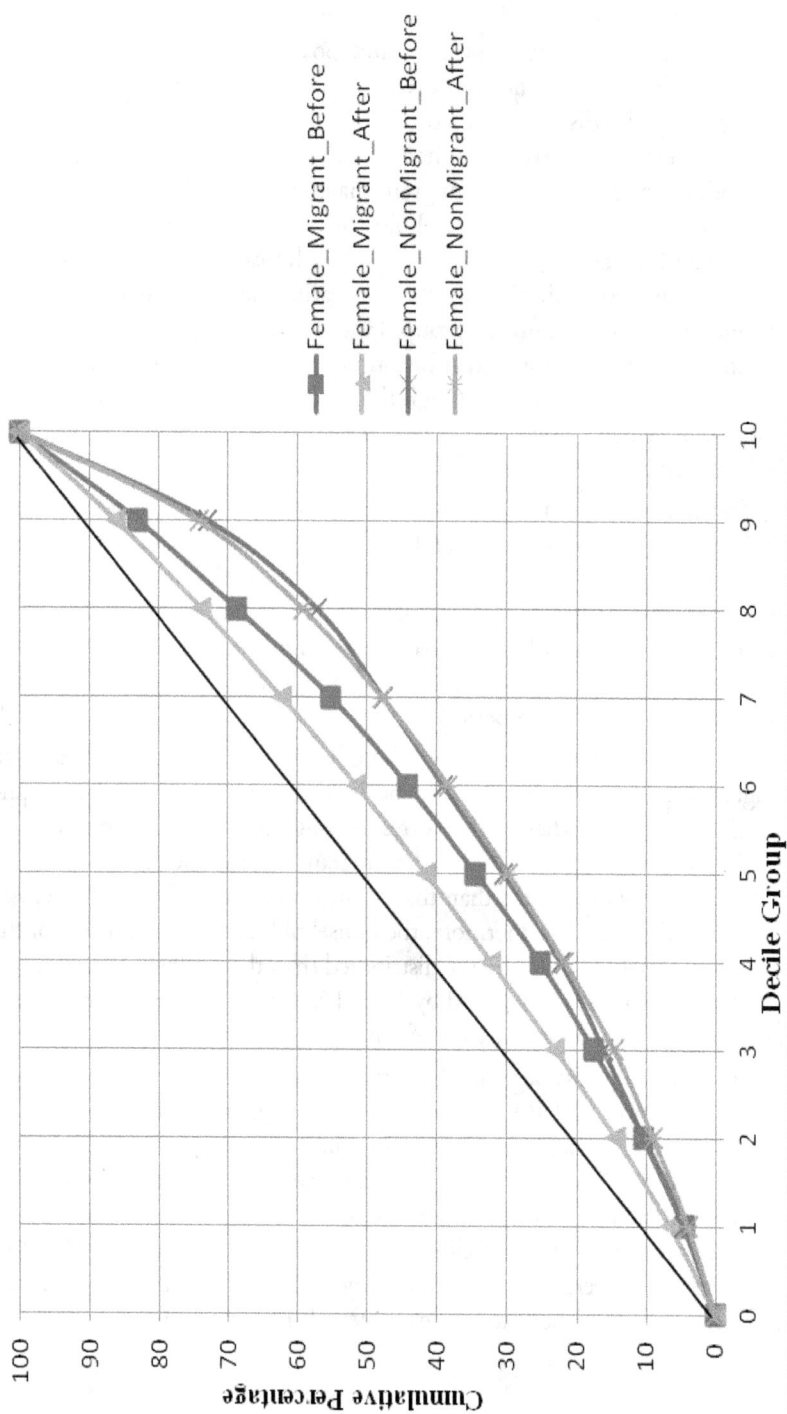

Figure 8.2: Female Migrants and Non-migrants

Figures 1 and 2 illustrate the income distribution of male and female respondents by migrant status, across the pre- and post-migration/interview periods. The central 450 diagonal line is the equi-income distribution curve, representing perfectly equally distributed income. Figure 1 shows that the household income in the earlier period was more unequally distributed than during the later period for both the migrant and non-migrant males. Between the two groups, the Lorenz curves depict the migrant households' income to be more equally distributed than that of the non-migrants. Further, their household income distribution after migration improves, indicating that migration has resulted in narrowing down income inequality within the group. However, at points where the Lorenz curves for the two periods intersect, it becomes difficult to confirm this. Nevertheless, there is an indication that although the income distribution of the non-migrants also improved in period two, there is a relatively greater improvement in the income distribution of the migrants.

In the case of females, a similar trend is observed with respect to the migrant and non-migrant groups in general. The income distribution is far more unequally distributed in period one for the non-migrants, which narrows down very little in period two. Whereas in both periods, the migrant households' incomes are far more equally distributed. Here again, the intersection of Lorenz curves of the non-migrant households makes it difficult to confirm the decline in income inequality across the two periods.

In effect, Figures 1 and 2 show the Lorenz curves for male migrants and non-migrants, and female migrants to be lying relatively closer to the equi-income distribution curve, whereas the corresponding curves for female non-migrants lie farther away from it. This implies that migrant male and female household incomes are more evenly distributed than that of the non-migrants by gender across the two periods under study. Furthermore, the household income distribution of the female migrants is slightly more equally distributed than that of the migrant males in both the periods. This is also depicted by Table 10, containing the Gini index for the two groups by gender, across the two reference periods of study.

Income inequality among the female migrant households is observed to be higher in the pre (40.46) than in the post (35.80) migration period. The female non-migrant households have the highest income inequality across the two periods (46.27 and 45.92 respectively). Although the male migrants (44.33) had higher income inequality than the non-migrants (43.17) in period one, their post-migration income inequality (41.75) declined more than that of the latter (42.19) in period two. However, overall, income inequality had registered a decline across the two reference periods for both the male and female households by migrant status.

Table 8.10: Gini Coefficient

Sl. No.	Details	Male	Female
A.	**Migrant Household**		
	i) Before	44.33	40.46
	ii) After	41.75	35.80
B.	**Non-Migrant Household**		
	i) Before	43.17	46.27
	ii) After	42.19	45.05

Conclusion

The paper analysed the factors influencing migration decision by gender, and verified whether adverse climatic changes influence it in Puri district, Orissa. It also examined the changes in socio-economic conditions of the sample migrant and non-migrant respondents, and their income distribution over a period of three years before and after migration/interview. Furthermore, the pull and push factors of migration, problems, and the reasons for non-migration were examined. The study is based on primary data collected from a random sample of 120 migrant and non-migrant respondents each, comprising 60 males and 60 females each from four sample villages in the flood affected Rupdeipur Gram Panchayat of Pipli block of Puri district, from June-July 2009.

The findings reveal the migrant households to be comparatively socially and economically poorer than their non-migrant counterparts, while across gender the female households were relatively worse off than the male households. The majority of both categories of the sample households had reported being severely affected during the floods of June-September 2008 and the earlier ones.

As regards employment status, a vast majority of the migrants was seasonally and temporarily employed, whereas their non-migrant counterparts were either working on their own land or unemployed or housewives. Post-migration, the migrants were more regularly employed in their neighbouring state, while not much change was observed in the case of the non-migrants in the village.

An analysis of the decision to migrate revealed that while better education level and more land ownership negatively and significantly affected it regardless of gender, the influence of social network was positive and significant. The latter finding is confirmed by the studies Munshi 2003; Bauer et al. 2006 and Molaei, Santhapparaj and Malarvizhi 2008. On the other hand, while pre-migration household income level significantly discouraged female migration, it significantly encouraged male and combined migration. Studies of Ellis (2009), and Ezra and Kiros (2001) revealed similar income diversification effects.

Higher pre-migration standard of living was found to significantly reduce male and combined migration, whereas it significantly encouraged female migration. Meanwhile, an increase in pre-migration annual employment days in native village significantly discouraged female and combined respondent's migration, whereas a rise in the degree of climatic change significantly encouraged it. The latter is supported by the findings of Black et al. (2008), who reported increased migration to safer locations in response to aggravating climatic change.

Contrary to expectation, higher household debts were found to significantly discourage male migration, whereas higher dependency ratio significantly increased it. With respect to the combined respondents, marital status significantly encouraged migration.

The foremost push factor for both male and female migrants was poverty, caused by factors like lack of assets and climate change hazards (Afifi 2010). Meanwhile, better income and job opportunity constituted the main pull factors regardless of gender (Ellis 2009; and Ezra and Kiros 2001). Wide variations were observed in the priorities attached to reasons for non-migration and problems faced by the migrants by gender, which are attributable to the different socialisation norms prevailing in the society.

Regardless of gender, the impact of migration on the migrant households revealed significant increase in their labour supply per day/month/annum, income, household expenditure, savings, debt and SLI in the post-migration period, with the significance level emerging relatively higher for females in all cases except SLI. Whereas, comparatively greater reduction in income inequality was observed among the migrant than non-migrant sample households.

In sum, the flood affected villages of Rupdeipur Gram Panchayat in Pipli block of Puri district have compelled the low skilled rural poor to resort to migration as adaptation to climatic changes, survival and livelihood diversification strategies. Although multiple factors have led to migration in the study area, climate change and poverty have also been identified as its significant determinants. Moreover, the findings revealed evidences of gender variations in the factors leading to migration. The study calls for implementation of suitable adaptation, coping and support policies from a gender perspective. This would include provisions of alternative employment schemes to protect and mitigate the problems of the affected people, training and land management methods, in the selected villages. The problems of migration due to climate change can also be minimised through timely dissemination of information on impending climatic catastrophe with the help of Meteorological department and media to ensure preparedness of the villagers in advance that could minimise damage and loss of life and material. Towards this end, the National Disaster Management Authority (NDMA) in the State is engaged in activities to reduce the vulnerability of coastal areas to cyclones, with the involvement of Orissa State Disaster Management Authority (OSDMA) of NDMA. Its aim is to protect target villages during disasters, besides facilitating communication under natural calamities, in coordination and protection of the fishing communities. Information Education and communications (IEC) activities are conducted for Dissemination of Early Warning and Safety, through the media to inform villages of advancing catastrophe at least two days before the flood, to ensure preparedness for mitigation and safety of all people (Government of Orissa 2013). In addition, the Orissa Water Resources Department (WRD) under National Cyclone Risk Mitigation Project (NCRMP) is managing the saline embankments, besides having a Resettlement Action Plan (RAP) under the NCRMP; which is implemented under the guidance of environmental and social management framework (ESMF). However, the effective implementation of these ongoing government projects and schemes requires the committed, integrated and coordinated efforts of all the relevant stakeholders.

References

Afifi, T., 2010, 'Economic or Environmental Migration? The Push Factors in Niger', *International Migration*, Vol. 49, No. S1, pp. e95–e124.

Ahmed, S.A., Diffenbaugh, N.S., Hertel, T.W., Lobell, D.B., Ramankutty, N., Rios, A.R. and Rowhani, P., 2009, *Climate Volatility and Poverty Vulnerability in Tanzania*, Policy Research Working Paper 5117, Washington DC: World Bank.

Government of Orissa, 2009, *Annual Report on Natural Calamities 2008-09*, Special Relief Commissioner, Revenue and Disaster Management Department, Bhubaneswar: Government of Orissa Publication. (http://www.orissa.gov.in/disaster/src/ANNU-AL_REP_04-05/2008-09/Natural_Calamities_2008-09.pdf). 10 May 2014.

Government of Orissa, 2010, *Annual Report on Natural Calamities 2009 - 10*, Special Relief Commissioner, Revenue and Disaster Management Department, Bhubaneswar: Government of Orissa Publication. (http://www.orissa.gov.in/disaster/src/ANNU-AL_REP_04-05/Annual_Report_2009-10.pdf). 10 May 2014.

Asian Development Bank, 2009, *Climate Change and Migration in Asia and the Pacific*, Manila, Philippines: Asian Development Bank.

Barnett, J. and Adger, N., 2007, 'Climate Change, Human Security and Violent Conflict', *Political Geography*, Vol. 26, No. 6, pp. 639–655.

Barnett, J. and Webber, M., 2009, *Accommodating Migration to Promote Adaptation to Climate Change*, Commission on Climate Change and Development, Stockholm, Sweden: Commission on Climate Change and Development. (http://www.ccdcommission.org/Filer/documents/Accommodating%20Migration.pdfv). 12 May 2014.

Bauer, T., Epstein, G.S. and Gang, I.N., 2006, 'The Influence of Stocks and Flows on Migrants' Locational Choices', *Review of Development Economics*, Vol. 10, No. 4, pp. 652-65.

Black, R., Kwiveton, D., Skeldon, R., Coppard, D., Murata, A. and Schmidt-Verkerk, K., 2008, *Demographics and Climate Change: Future Trends and their Policy Implications for Migration*, Migration DRC Working Paper T-27, U.K.: University of Sussex.

Black, R., Adger, W.N., Arnell, N., Dercon, S., Geddes, A. and Thomas, D.S.G., 2011, 'The Effect of Environmental Change on Human Migration', *Global Environmental Change*, Vol. 21, Supplement 1, pp. S3-S11.

Boano, C., Zetter, R. and Morris, T., 2008, *Environmentally Displaced People: Understanding the Linkages between Environment Change, Livelihoods and Forced Migration*, Forced Migration Policy Briefing Refugee Studies Centre, United Kingdom: University of Oxford.

Casillas, C.E. and Kammen, D.M., 2010, 'The Energy-Poverty-Climate Nexus', *Science*, Vol. 330, November, pp. 1181-1182.

Chau, N., 1997, 'The Pattern of Migration with Variable Migration Costs', *Journal of Regional Science*, Vol. 37, No. 1, pp. 35-54.

Doevenspeck, M., 2008, *The Fine Line between Choice and Flight: Environmental Drivers, Socio-Economic Processes and the Perpetuation of Migration in Rural Benin, West Africa*, Paper Presented at the International Conference on Environment, Forced Migration and Social Vulnerability (EFMSV), Bonn, Germany, 9–11 October.

Durlauf, S., 2001, 'A framework for the Study of Individual Behaviour and Social Interactions', *Sociological Methodology*, Vol. 31, pp. 47-87.

Easterling, W.E., Aggarwal, P.K., Batima, P., Brander, K.M., Erda, L., Howden, S.M., Kirilenko, A., Morton, J., Soussana, J.F., Schmidhuber, J. and Tubiello, F.N., 2007, 'Food, Fibre and Forest Products', in M.L. Parry, O.F. Canziani, J.P. Palutikof, P.J. van der Linden and C.E. Hanson, eds., *Climate Change 2007: Impacts, Adaptation and Vulnerability, Contribution of Working Group II to the Fourth Assessment Report of the Intergovernmental Panel on Climate Change*, Cambridge, UK: Cambridge University Press.

Ellis, F., 2000, *Rural Livelihoods and Diversity in Developing Countries*, Oxford: Oxford University Press.

Government of Orissa, 2013, *Environmental and Social Screening from National Cyclone Risk Mitigation Project*, Document 2009105/EC/Resettlement Action Plan (RAP), Environment and Ecology Department, Orissa: Government of Orissa Publication.

Epstein, G.S. and Gang, I.N., 2004, *The Influence of Others on Migration Plans*, IZA Working Paper 1244, Bonn, Germany: IZA.

Ezra, M. and Kiros, G.E., 2001, 'Rural Out-Migration in the Drought Prone Areas of Ethiopia: A Multilevel Analysis', *International Migration Review*, Vol. 35, No. 3, pp. 749–771.

Findley, S.E., 1994, 'Does Drought Increase Migration? A Study of Migration from Rural Mali during the 1983-1985 Drought', *International Migration Review*, Vol. 28, No. 3, pp. 539–553.

Garret, Henry E. and Woodworth, R.S., 1969, *Statistics in Psychology and Education*, Bombay: Vakils, Feffer and Simons Private Limited.

Glaeser, E.L. and Scheinkman, J.A., 2001, 'Measuring Social Interactions', in S.N. Durlauf and H.P. Young, eds., *Social Dynamics*, Washington, D.C.: Brooking Institute, pp. 83-131.

Gómez, Oscar, 2013, *Climate Change and Migration: A Review of the Literature*, A Study Commissioned by the International Institute of Social Studies, Rotterdam, The Hague: Erasmus University.

Gray, C. and Mueller, V., 2012, 'Drought and Population Mobility in Rural Ethiopia', *World Development*, Vol. 40, No. 1, pp. 134-145.

Groen, J.A., and Polivka, A.E., 2009, *Going Home after Hurricane Katrina: Determinants of Return Migration and Changes in Affected Areas*, Working Paper 428, Washington, D.C.: US Bureau of Labor Statistics.

Guterres, A., 2008a, 'Millions Uprooted: Saving Refugees and the Displaced', *Foreign Affairs*, Vol. 87, No. 5, pp. 90–99.

Guterres, A., 2008b, *Climate Change, Natural Disasters and Human Displacement: A UNHCR Perspective*, Geneva: United Nations High Commissioner for Refugees. (http://www.un-hcr.org/refworld/type, RESEARCH,UNHCR,492bb6b92,0.html.). 12 May 2014.

Helmenstein, C. and Yegorov, Y., 2000, 'The Dynamics of Migration in the Presence of Chains', *Journal of Economic Dynamics and Control*, Vol. 24, No. 2, pp. 307-23.

Henry, S., Schoumaker, B. and Beauchemin, C., 2004, 'The Impact of Rainfall on the First Out-migration: A Multi-Level Event-History Analysis in Burkina Faso', *Population and Environment*, Vol. 25, No. 5, pp. 423–460.

Horton, G., Hanna, L. and Kelly, B., 2010, 'Drought, Drying and Climate Change: Emerging Health Issues for Ageing Australians in Rural Areas', *Australasian Journal on Ageing*, Vol. 29, No. 1, pp. 2-7.

Huisman, H., 2005, 'Contextualising Chronic Exclusion: Female-Headed Households in Semi-Arid Zimbabwe', *Tijdschrift Voor Economische En Sociale Geografie*, Vol. 96, No. 3, pp. 253-263.

McIntyre, B.D., Herren, H.R., Wakhungu, J. and Watson, R.T., eds., 2009, *Agriculture at Crossroads, Global Report*, International Assessment of Agricultural Knowledge - IAASTD, Science and Technology for Development, Washington, D.C., USA: Island Press.

International Organisation for Migration, 2007, *Discussion Note: Migration and the Environment* (MC/INF/288 – 1 November 2007 - Ninety Fourth Session), 14 February 2008IOM, Geneva, Switzerland: International Organization for Migration.

Kartiki, Katha, 2011, 'Climate Change and Migration: A Case Study from Rural Bangladesh', *Gender and Development*, Vol. 19, No. 1, March, pp. 23-28.

Laczko, F. and Aghazarm, Ch., eds., 2009, *Migration, Environemt and Climate Change: Assessing the Evidence*, Geneva: IOM.

Manski, C., 2000, 'Economics Analysis of Social Interactions', *Journal of Economic Perspectives*, Vol. 14, No. 3, pp. 115-36.

Molaei, Mohammed Ali, Santhapparaj, A. Solucis and Malarvizhi, C.A., 2008, 'Rural-Urban Migration and Earning Gains in Iran', *Journal of Social Sciences*, Vol. 4, No. 3, pp.158-164.

Munshi, K., 2003, 'Networks in the Modern Economy: Mexican Migrants in the US Labor Market', *Quarterly Journal of Economics*, Vol. 118, No. 2, pp. 549-99.

Naik, A. Stigter, E. and Laczko, F., 2007, *Migration, Development and Natural Disasters: Insights from the Indian Ocean Tsunami*, IOM Migration Research Series No. 30, Geneva: International Organization for Migration.

Nelson, V., Meadows, K., Cannon, T., Morton, J. and Martin, A., 2002, 'Uncertain Predictions, Invisible impacts, and the Need to Mainstream Gender in Climate Change Adaptations', *Gender and Development*, Vol. 10, No. 2, pp. 51-59.

Nelson, G.C.R., Rosegrant, Mark W., Palazzo, Amanda, Gray, Ian, Ingersoll, Christina, Robertson, Richard, Tokgoz, Simla, Zhu, T., Sulser, T.B., Ringler, C., Msangi, S. and You, L., 2010, *Food Security, Farming and Climate Change to 2050: Scenarios, Results, Policy Options*, IFPRI Research Monograph, Washington DC: IFPRI e-brary. (ebrary. ifpri.org/cdm/ref/collection/p15738coll2/id/127066). 10 May 2015.

Office of Registrar General and Commissioner, Various Years, *Census*, New Delhi: Government of India.

Omolo, N., 2011, 'Gender and Climate Change-Induced Conflict in Pastoral Communities: Case Study of Turkana in North-western Kenya', *African Journal on Conflict Resolution*, Vol. 10, No. 2, pp. 81-102.

Planning Commission, 2012, *Press Note on Poverty Estimates: 2009-10*, New Delhi: Government of India, 19 March. (http://planningcommission.nic.in/news/press_pov1903.pdf.). 10 May 2014.

Radu, Dragos, 2008, 'Social Interactions in Economic Models of Migration: A Review and Appraisal', *Journal of Ethnic and Migration Studies*, Vol. 34, No. 4, May, pp. 531-548.

Renaud, Fabrice G., Bogardi, Janos J., Dun, Olivia and Warner, Koko, 2007, *Control, Adapt or Flee: How to Face Environmental Migration?*, InterSecTions, No. 5/2007, Bonn, Germany: UNU-EHS.

Renaud, Fabrice G., Dun, Olivia, Warner, Koko and Bogardi, Janos, 2011, 'A Decision Framework for Environmentally Induced Migration', *International Migration*, Vol. 49, Supplement 1, pp. e5-e29.

Roy, T.K., Jayachandran, V. and Banerjee, Sushanta K., 1999, 'Economic Condition and Fertility: Is there a Relationship', *Economic and Political Weekly*, Vol. 34, Nos. 42s & 43, October 16-23, pp. 3041- 3046.

Ruel, M.T., Garrett, J.L., Hawkes, C. and Cohen, M.J., 2010, 'The Food, Fuel, and Financial Crises Affect the Urban and Rural Poor Disproportionately: A Review of the Evidence', *Journal of Nutrition*, Vol. 140, No. 1, January, pp. 170S-6S.

Stal, M., and Warner, K., 2009, *The Way Forward: Researching the Environment and Migration Nexus*, UNU-EHS Research Brief, Bonn, Germany: Institute for Environment and Human Security, United Nations University.

Tacoli, Cecilia, 2009, 'Crisis or Adaptation? Migration and Climate Change in a Context of High Mobility', *Environment and Urbanization*, Vol. 21, No. 2, pp. 513-525.

Terminski, Bogumil, 2012, Environmentally-Induced Displacement: Theoretical Frameworks and Current Challenges, May, Geneva.

Thadani, V.N. and Todaro, M.P., 1984, 'Female Migration: A Conceptual Framework', in J.T. Fawcett, S.E. Khoo and P.C. Smith, eds., *Women in the Cities of Asia: Migration and Urban Adaptation*, , Boulder, Colorado, United States: Westview Press, pp. 36-59.

United Nations Development Programme, 2009, Overcoming Barriers: Human Mobility and Development, *Human Development Report 2009*, New York: Palgrave Macmillan.

Vincent, K., Cull, T. and Archer, E., 2010, 'Gendered Vulnerability to Climate Change in Limpopo Province, South Africa', in I. Dankelman, ed., *Gender and Climate Change: An Introduction*, London: Earthscan, pp. 160-167.

9

Effects of Climate Change and Heat Waves on Low Income Urban Workers: Evidence from India[1]

Saudamini Das

Introduction

A heat wave is an extended period of hot and humid weather measured relative to the normal weather pattern of an area (Meehl and Tebaldi 2004). The frequency and duration of heat waves have gone up with climate change, and the world is facing hot days, hot nights, and heat waves more frequently (IPCC 2007a). Heat waves are a health hazard as they slow the evaporation of perspiration, which cools the human body. The human skin temperature is strongly regulated to remain at 350 C or below, under normal conditions (Sherwood and Huber 2009). The skin temperature has to be lower than the core body temperature (370 C) for metabolic heat to be transmitted to the skin. Sustained skin temperature above 350 C due to heat waves elevates the core body temperature, causes tiredness, nausea, body ache, etc., and thus, reduces the person's work efficiency.[2] High temperatures can pose serious threats to all individuals, not necessarily the old and the unwell; hence every possible effort should be made to maintain one's skin temperature at around 350 C (Mehnert *et al.* 2000; Bynum *et al.* 1978). Heat waves are being experienced with increasing frequency in different parts of the world (Das and Smith 2012; IPCC 2007b, 2014) and people are resorting to various adaptive measures to reduce the adverse effect on health. The present paper discusses some such measures adopted by poor urban informal sector workers and accounts for the loss suffered and costs incurred by them in some parts of India.

Heat Waves and Urban Labour Market

There is seasonality in the rural work pattern, which may help rural workers by reducing their exposure to extreme heat during heat waves. But urban workers in secondary or tertiary sector activities – which are little influenced by weather, continue yearlong, and occur in poor and exposed environments – are more vulnerable to weather-related hazards like heat waves that continue for a few days. Other than the health implications on the working poor, such climatic changes may impact labour supply and productivity, particularly in labour-intensive activities.

There is much discussion in the labour economics literature on the labour-leisure model of unemployment. Using micro-data, the inter-temporal labour-leisure elasticity of substitution is estimated to be 0.2 (McCurdy 1981). Labour substitutes labour for leisure usually under external stress, like the presence of consumption commitments (for example, expenditure associated with children such as food and schooling). Short-run wage declines also motivate workers to increase short-run market hours to maintain cash flow (Dau Schmidt 1984). Now, the question arises: how is this income induced labour-leisure substitution being affected by exogenous stress from heat waves, which may impose multiple constraints on the worker (like low-income due to both less availability of work and low productivity due to heat effect on health, high expenditure to adapt to extreme heat, etc.)?

During the peak hot hours of heat wave days, workers may withdraw from the market, or there may be fewer customers than usual; therefore, workers may earn less. To offset this loss, workers may like to work longer, but may not be able to do this if the type of work provides little scope for such substitution,[3] or the thermal stress is strong enough to force him/her to rest more. Poor health reduces the capacity to work and has significant effects on wages (Currie and Madrian 1999). In this light, one may presume that heat-affected workers may not be able to work longer, and, therefore, might earn less. If income recovery is not possible, then the person is forced to remain at a lower level of wellbeing. The loss of income and the additional expenditure, if any, to cope with the extreme heat can be a measure of private adaptation cost, especially for self-employed people or informal sector workers. Many of the developing countries, especially in South Asia, are poised for faster growth, which involves large-scale construction and other exposure-based activities. If the present trends in temperature continue, this may involve large increases in private and public expenditure on adaptation, more provision of electricity, change in technology and work environment, etc. There is little research on these issues, though the research findings are likely to have strong policy implications.

Focus of the Study

In this paper, an attempt is made to find out the burden in terms of working hour loss, loss in family work time, and monetary cost incurred by low-income urban informal workers because of severe heat waves in two cities of Eastern Indian state of Odisha – Bhubaneswar and Sambalpur. These two cities, along with most of the rest of the state, have been witnessing severe heat waves since 1998, and both the people and the government have been adopting various strategies to cope.[4] The heat wave management efforts of the government have recorded some success in terms of reducing mortality (Das and Smith 2012; Das 2014). Based on daily maximum temperature recorded with the local meteorology stations, Das and Smith (2012) measured the yearly heat wave days for different areas of the state, including these two cities, for the period 1998 to 2012 (Figure 1). Clearly, heat wave days are increasing over years and, on average, it calculates 12 heat wave days per year for these two cities.

Figure 9.1: Annual Counts of Heat Wave Days in Bhubaneswar and Sambalpur, Odisha

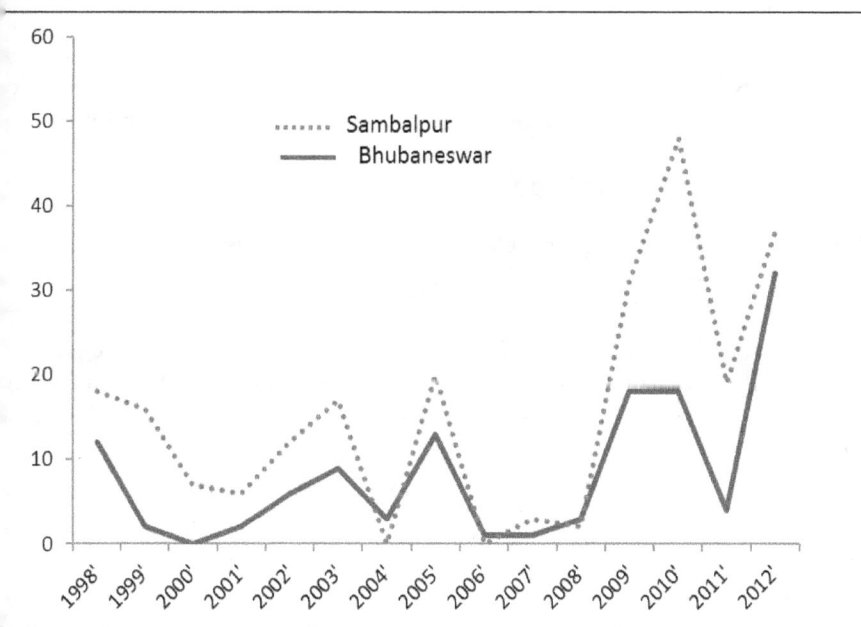

Source: Das and Smith, 2012.

Most workers in this study who are exposed to heat are in a sense, own-account enterprises in trade, transport and services activities and who, unlike the service class, make their own independent labour allocation decision to maximise utility. Gronau (1987) provides an excellent review of studies on home production. Work at home is said to be a close substitute to work in the market in terms of the direct utility these activities generate. However, to understand these issues at the household level, time budget data are required, which this study attempts to collect through a questionnaire survey.

A primary survey based on purposive random sampling was pursued in these areas, to get responses to the questionnaire. An attempt was made to find out the time allocation of people to different activities during a day (7 am to 11 pm) and other queries like if people suffered from any type of heat attack, changed their occupation because of heat, know of the government's heat wave awareness campaign, their expenditure to cope with extreme hot temperature, and so on. This is an exploratory study to find out the prevalence of the problem and its effect, if any, on labour supply. Therefore, most questions on labour allocation were phrased to be indirect, such as 'What do you do during 7 am to 9 am now (i.e. yesterday)?' and not 'What do you do during 7 am to 9 am during a heat wave day?' The interview was conducted during peak summer (not heat wave), and the area had experienced a heat wave two weeks before, so incorrect reporting for heat wave days are supposedly low. The sample survey section describes in detail the procedure followed to collect the time allocation data.

The sections below present some theoretical literature on heat stress and labour supply issues followed by the sample survey, the descriptive statistics of the sample, the results and then the adaptation cost estimates. The last section concludes.

Heat Stress and Labour Supply

The effect of climate change on labour supply has been discussed in terms of labour availability constraints in vulnerable regions because of the migration of labour from vulnerable areas to less vulnerable areas.[5] Regular heat waves may have no effect on total labour availability in a region; as such phenomena do not induce migration, but are likely to constrain labour productivity by restricting the individual's ability to work efficiently due to unbearable weather. Heat stress may alter the marginal productivity of labour or the marginal cost of supplying labour to activities where the individual is exposed to heat. Thus, under a heat wave scenario, one expects change in workers' decision regarding the allocation of time from labour to rest, especially in exposed sectors, such as agriculture, construction, manufacturing, etc. (Zivin and Neidell 2010). Studies show that marginal productivity of labour gets impacted by lower endurance, fatigue (Gonzalez-Alonso et al. 1999; Galloway and Maughan 1997; Nielsen et al. 1993), and cognitive performance (Epstein et al. 1980; Ramsey 1995; Hancock et al. 2007; Pilcher et al. 2002), etc., and heat stress can cause all such health effects. Studies under ergonomics have also established a strong association between productivity loss and rise in temperature (Niemelä et al. 2002; Lan et. al. 2010; Mahamed and Srinavin 2005).[6] Simultaneously, the marginal utility from leisure activities may go up during peak heat (Ma et al. 2006; Pivarnik et al. 2003; US Department of Health and Human Services 1996; Eisenberg and Okeke 2008) triggering the leisure-work substitution. This may have serious welfare implication for people, especially for low-income workers for whom every hour of work matters. The link between external shock from heat waves and the wellbeing of poor labour class is yet to be explored in detail, though there are plenty of studies on the effect of heat waves on human health, especially in epidemiology (Kovats and Hajat 2008). Based on an aggregative analysis on temperature and labour supply at the level of counties, a careful analysis by Zivin and Neidell (2010) showed large reductions in labour supply in climate-exposed industries in the US as the temperature rose over 85° F. Secondly, at higher temperatures, unemployed people reduced outdoor leisure activity; this shows their preference to stay indoors during heat waves. Effective labour supply – defined as a composite of labour hours, task performance, and effort – is found to decrease during temperature deviations from the biological optimum, according to a study that uses country-level panel data on population-weighted average temperature and income (1950–2005) to illustrate the potential magnitude of this effect (Heal and Park 2014). The present paper uses individual level questionnaires survey data and is based on a developing region where the effect could be much stronger. Thus, the present findings are likely to provide complementarities to the evidence found from developed economies.

Beyond labour-leisure substitution, another setback could be a longer absence from home on heat wave days. This could take one of the three forms: (1) leave home earlier than usual; (2) return home late to avoid travelling in the scorching sun; or (3) work overtime in the evening or at night to recoup the loss in income during the usual working hours. This means being unable to help in housework. Such absence may be very high, especially for those who live far from work, and their family members may suffer more compared to others because of heat waves. The effect of heat waves on family work is less talked about, as it is usually considered a part of leisure activities in labour allocation studies. This study accounts for the time explicitly allocated to house. The study focuses on very poor self-employed workers whose family do all household related works manually because of low income and family members help each other. If a member is away from home for long, other members of the household are likely to be over-burdened. While losing time for housework may not reduce income, it may have utility implications, as the utility from housework is different from the utility from rest.

The labour class reallocates time between labour and leisure to maximise welfare because of the income effect. During heat waves, a reversal of such allocation may occur because either leisure provides more utility, or because the likely health effect of working long hours under heat wave conditions will be too adverse, and require a lot of medical expenditure, which may have serious economic implications. Therefore, during heat waves, workers allocate time to different activities to minimise the negative impacts of heat stress on health. Thus, the choice of activity and location, i.e. whether to work at home or at work (place) or spend time under a shady tree (rest), as happens in underdeveloped countries, is undertaken rationally to minimise the negative impact of temperature by reducing the exposure to hot weather. I take the hypothesis that workers reallocate time between work, rest and time spent on housework to minimise the negative impact of heat waves on health.

Any marginal change in time required for rest will impact the time allocated to housework or outside work depending on the seriousness of the heat wave. The income effect of outside work being very strong for poor people, the individual will first reallocate time from housework to rest, but ultimately may reduce outside work time, as the scope of reducing housework may not be great. As, usual, housework takes a few hours every day, and some of it (like cooking or helping in cooking, buying groceries, etc.) is important and cannot be ignored, the individual will be forced to allocate more time from economic work to rest if heat stress requires more rest. On a heat wave day, the primary objective of the worker is to minimise the negative health effect of the heat wave; therefore, rest assumes utmost importance, and work at home and outdoors is adjusted to ensure enough rest. The temperature anomaly determines the hours of rest needed and thus determines the labour allocation decision on the heat wave day.

Sampling and Survey

A purposive random sampling was conducted in Bhubaneswar and Sambalpur – two cities in Odisha affected repeatedly by heat waves – to collect information on adaptation to heat stress by poor urban workers. Ten types of urban workers from the low economic strata who work mostly in the open environment were chosen: vegetable/fruit sellers, cobbler, construction workers, porters, rickshaw/trolley driver, auto rickshaw driver, taxi driver, mobile marketing and sales executive/ representative, vendors (mobile sellers of household items in trolleys), and owner and workers in open-air retail enterprises (temporary stall owners).

The sample was taken from comparatively crowded and backward market areas of the cities, like areas close to railway station, bus stop etc., as workers in these areas are poorer. The sample was drawn randomly by picking 15 workers from each of these 10 categories. The questionnaire-based survey was conducted simultaneously in both cities in 2013 between 25 April and 20 May, when the temperature was around 42–430 C in Sambalpur and 400 C in Bhubaneswar, but it was not a heat wave period. It is considered a heat wave when the temperature is around 450 C in Sambalpur and 420 C in Bhubaneswar, which occurs many times during the summer and occurred around 15 April 2013 also. As reported before, answers to most of the questions on time allocation to different activities for a heat wave day were elicited through recall. The temperature was already high, and people were asked what they did in each two-hour period (7–9 am, 9–11 am, etc. going up to 9–11 pm) the previous day and what they did on the heat wave day (using the local name for heat waves) they experienced around 15th April that year; therefore, it is hoped that people's answers were mostly accurate. The survey listed workers' activities during each two-hour period between 7 am and 11 pm for these two days (the day before the interview date (a normal summer day) and the heat wave day two weeks before), and then grouped the activities into 'rest', 'house work', and 'outside work'. Next, the hours spent on each of these three groups of activity were counted to measure the time allocation on a normal summer day and on a heat wave day. Along with time allocation, work related and biographical information, the survey also collected information on purchase of consumer durables and other expenditures to cope with heat stress and differences in regular day-to-day expenditures between heat wave days and normal summer days.

Sample Features

The sample comprised mainly middle-aged people and was dominated by men. In the study area, occupations surveyed are usually undertaken by males, so the sample had just 6 per cent female respondents. There was a mix of different education levels in the sample; most respondents were educated up to Class 10. Most of the uneducated respondents were either cobblers, or rickshaw drivers, or retail sellers; whereas mobile marketing and sales executives were graduates with one or two degrees. Only 4 per cent of the sample had some technical education like motor garage work, driving, electrical training, and so on.

On social class distribution, 18 per cent belonged to the general class, 43 per cent were backward classes, 30 per cent belonged to the scheduled caste, and 9 per cent to scheduled tribes. The family size was 5.17 members on average, although some families had as many as 18 members. On average, most families had 3.5 dependents. Most respondents were migrants; only 29.3 per cent were born in these cities. Around 75 per cent of the respondents were household heads.

Table 1 shows the distribution of different occupation class in the sample based on annual income range. The modal annual income of respondents was between INR 50,000[7] and INR 100,000, which includes the incomes of 52 per cent of the respondents. Some were very poor (1 per cent), with an annual income of less than INR 10,000, and some were relatively well-off (1 per cent), with an annual income of more than INR 300,000.

The cobblers were the poorest – 45 per cent made less than INR 20,000 annually – followed by porters or manual labourers in transport and other sectors. Mobile marketing and sales executives made INR 50,000–100,000 annually at the least and were the richest in the sample; 80 per cent of them made INR100, 000–300,000 annually.

Health Problems from Heat Waves

Some workers reported that heat waves caused them many health problems, like fever (15 per cent), tiredness (12 per cent), respiratory problems (8 per cent), losing consciousness (5 per cent), blurred vision (8 per cent), feeling of nausea (10.3 per cent), and body ache (4 per cent), etc., though most of them had no prior history of health problems. Of the respondents who reported health problems, most were cobblers, porters, rickshaw pullers, or auto drivers; there were few marketing and sales executives. Hospitalisation due to heat attack was reported by 15 per cent of workers, of which 50 per cent were either porters (18 per cent), or rickshaw pullers (16 per cent) or cobblers (13 per cent), and the rest from other categories. This clearly indicates that the major victims of heat attack are in the lowest income categories having more exposed occupations from among the low-income workers. Only 2.3 per cent of the respondents reported changing their occupation, in fact their mode of working, because of heat. They reported to have changed their work from mobile vending to stall vending at roadside like 'from selling eggs or other products at door step to selling tea or sugar cane juice at a fixed place'.

Table 9.1: Distribution (per cent) of the Sample According to Occupation and Income Class

Occupation	Annual Income Class (in INR)					
	Less than 10000	10000 - 20000	20000 - 50000	50000 - 100000	100000 - 30000	Above 300000
Vegetable/Fruit Seller	0.03	0	0.43	0.47	0.07	0
Cobbler	0.10	0.35	0.13	0.29	0.13	0
Construction Worker	0	0.03	0.63	0.33	0	0
Porters (Manual Labour in Transport or Other sectors)	0	0	0.38	0.62	0	0
Rickshaw/Trolley Drivers	0	0	0.53	0.43	0.03	0
Auto Driver	0	0	0.13	0.55	0.32	0
Taxi Driver	0	0	0.13	0.73	0.13	0
Mobile marketing and Sales executive	0	0	0	0.13	0.80	0.07
Vendors (Mobile sellers of HH items in trolleys)	0	0	0.04	0.93	0.04	0
Owners & workers in open retail enterprises	0	0.06	0	0.74	0.19	0
Total	0.01	0.05	0.24	0.52	0.17	0.01

Effect of Awareness Programmes on People

Since 2003, the Government of Odisha has been conducting an awareness programme on what to do during heat waves and the things to avoid (see Appendix for detail). When asked if they knew of the programme, find it useful, and if they have taken the advice, around 99 per cent said they knew of the programme. The source of information was radio (63 per cent), newspaper (64 per cent), television (73 per cent), pamphlets (31 per cent), volunteers (18 per cent), neighbours (14 per cent), and NGOs (12 per cent), etc. Most had received the information from multiple sources; 83 per cent from maximum three sources[8] (either radio, or television, or newspaper or pamphlets), 13 per cent from two sources and only 4 per cent from only one source.

Around 93 per cent of the respondents found the government campaign very helpful and 99 per cent reported to have changed some of their habits during heat waves because of the campaign. On average, it was found that during heat waves, 73 per cent of respondents drink water and 65 per cent eat cucumber and onion before leaving home; 63 per cent carry a water bottle; 64 per cent carry an umbrella, 26 per cent wear light-colour cotton clothes; 21 per cent do not walk barefoot; and 33 per cent do not work during noon hours. Though the behavioural changes because of the campaign seem high, one does find a lower response rate for the ones involving more expenditure. It may be that poor economic condition restricts people from buying cotton clothes, walking slippers, or skip work during noon. Thus, the sample seemed to be a mixture of people who had experienced and suffered heat waves and who had taken some precautions to avoid health problems from subsequent experience due to awareness campaign. The impact of heat waves on labour supply of this informed and adapted working class is net of adaptation and thus, should be more useful for policymaking.

Results

Summary Statistics on Time Allocation

Workers allocate time differently on a heat wave day than on a normal summer day, and because of heat stress, they lose time (Table 2). It shows that workers have rested longer and spent less time on both outdoors work and housework, which, perhaps, is the first finding from the real life experience of poor workers in a very poor and deprived part of the world. On average, 1.19 hours of work time and 0.46 hours of family time were lost per heat wave day, as workers reallocated these hours to rest because of heat stress.

Table 9.2: Change in Time Allocation and Work Time Loss on Low-income Category Working Class Because of Heat Waves

Occupation	Change in average time allocation during a heat wave day compared to normal summer day (in hours)		
	Regular outside work	House work	Rest
Vegetable/Fruit Seller	-0.73	-0.22	0.95
Cobbler	-1.23	-0.47	1.70
Construction Worker	-1.54	-0.38	1.93
Porters (Manual Labour in Transport or Other sectors)	-1.62	-1.02	2.64
Rickshaw/Trolley Drivers	-1.65	-0.52	2.17
Auto Driver	-1.06	-0.21	1.28
Taxi Driver	-1.52	-0.68	2.2
Mobile marketing and Sales executive	-0.74	-0.37	1.1

Vendors (Mobile sellers of Household items in trolleys)	-1.26	-0.18	1.44
Owners & workers in open retail enterprises (Temp Stall)	-053	-0.58	1.11
Average change	-1.19	-0.46	1.65

This has very serious implications for poor workers who live from hand to mouth. Rickshaw drivers lost the most work time (1.65 hours), followed by porters (1.62 hours), construction workers (1.54 hours), taxi drivers (1.52 hours), etc., but the time lost is less for temporary stall owners (0.53 hours), vegetable sellers (0.73 hours), and sales executives (0.74 hours). This could be because temporary stall owners and vegetable sellers work either in the morning or late in the afternoon, and sales executives, who are comparatively well off, probably take better precautions so that they do not lose more work time. As expected, time loss from housework is low, as these activities require some minimum amount of time and such time required cannot be changed much, as explained before.

One may expect these people to compensate such loss by working longer other days, but the possibility of such adjustments are low due to many reasons. First of all, this extra rest is to neutralise the negative effect of heat on human body and these people have no luxury at home to recover quickly from such stress. Many reported 'not being able to sleep at night', 'feeling tired', 'irritated', 'having low motivation to work due to stress' etc. as other side effects of heat waves. So, over-exerting, as they all do manual job, to compensate for this loss of work time is difficult.

Cost of Adaptation

Next, private cost of adaptation is measured as income loss due to work time loss plus extra expenditures made to cope with the heat waves. As discussed above almost every respondent was aware of the awareness programme of the government and many had reported to have changed their behaviour, food, etc. and thus loss of work time is after the possible adaptations that each worker has undertaken. So loss of income from work time loss was added to extra expenditures. Extra expenditures information was collected on two heads: (i) on regular items like food, electricity, water, transport (either own or children's transport to school) etc. and (ii) on consumer durables like purchase of ceiling fan, water coolers, etc. made to cope with high temperature.

Expenditures on Non-durable Regular Items

Nobody could recollect the exact extra amount spent on regular items on heat waves days, but could report the approximate increase in expenditures on these heads in the month when heat waves were experienced. Hence using monthly expenditure differences, I calculate increase in expenditure on consumer non-durable goods per heat wave day. Table 3 shows these extra expenditures for different worker categories. All categories of workers are spending more on regular items in months with heat wave days compared to months with no heat waves, the average extra expenditure being INR 600, varying in between INR339 to INR810 for different groups. These extra expenditures constitute 4 to 15 per cent of the monthly income for different category of workers who earn approximately INR 85833 per annum or INR 7152 per month (approximately US$119). Spending 15 per cent of income by some workers to cope to heat waves in a month is too high a burden. As heat waves are felt during two months (mid-April to mid-June) in Odisha, the average numbers of heat wave days in a month are taken to be 6 (see section 1.2 above) and this calculates the extra expenditure on consumer non-durable goods to be INR 100 per heat wave day, which is 1.4 per cent of the monthly income.

Table 9.3: Economic Burden in Terms of Extra Monthly Expenditure on Food and Other Routine Items by Low-income Working Class Because of Heat Waves

Occupation	Average approximate annual income (in INR)a	Average increase in monthly routine expenditure due to heat wave days (in INR)b
Vegetable/Fruit Seller	63667	810 (0.15)
Cobbler	57903	597 (0.12)
Construction Worker	47667	450 (0.11)
Coolie (Manual Labour in Transport or Other sectors)	59828	597 (0.12)
Rickshaw/Trolley Drivers	57833	557 (0.12)
Auto Driver	110161	694 (0.08)
Taxi Driver	86333	612 (0.09)
Mobile marketing and Sales executive	200000	693 (0.04)
Vendors (Mobile sellers of Household items in trolleys)	78036	339 (0.05)
Owners & workers in open retail enterprises (Temp Stall)	95323	639 (0.08)
Average for all groups	85833	601 (0.08)
Average extra expenditure per heat wave day		100 (0.014)

a. Annual income was measured from mid values of income categories to which each worker belonged to.

b. Figures in parenthesis are the proportion of monthly income.

Expenditures on Consumer Durable Items

Regarding purchase of consumer durables, I use those purchases as adaptive expenditures where the respondent reported regular heat waves as the most important reason for such purchase.[9] Moreover, the source of money was stated as 'saving plus borrowing' by almost everybody who made such purchases due to heat waves and the year of purchase was either 2006 or afterwards. After 1998 heat waves, the state witnessed very intense heat waves during 2005 when more than 500 people died (Das and Smith 2012 and Figure 1) and this incident seems to have triggered these purchases. The costs of these durable consumer goods were annualised using the equivalent annual cost (EAC) method as shown below:

$$Annual\ Cost = \frac{P_i}{\frac{1-[\frac{1}{(1+r)^n}]}{r}}$$

where P_i is the price of the ith consumer durable, n is the possible life span of the commodity and r is the rate of interest which is the cost of borrowing. I used 8 per cent rate of interest to annualise the expenditures, though it could be much higher as people could be borrowing money from private sources. Thus, the annualised costs are the underestimates of actual costs. The information on life spans of these goods were also asked from the workers, but these information for some of the globally used goods like ceiling fans, refrigerators etc. also match the standard life expectancy chart of household goods reported by different home inspection experts.[10] Table 4 shows these calculations. As the proportion of the sample undertaking such activities varied widely from activity to activity, a weighted average of the annualised average cost was calculated using sample proportions as the weights to get this estimate for a representative worker.

Table 4 shows the annual averted expenditure on consumer durables to be INR 592 per sample household in a year, which works out to INR 49 per heat wave day. Adding extra expenditure on both durable and non-durables, the adaptation cost calculates to INR 149 per heat wave day or INR 1788 per year, as there are approximately 12 heat wave days in a year in the state. Using the approximate annual income and the number of hours worked during normal summer, the average hourly income was calculated to be INR 38 and loss of income per heat wave day as INR 46. Thus, the private adaptation cost of a low-income informal sector worker measures to INR 195 per heat wave day or INR 2340 per annum of which 51 per cent consists of extra expenditure on food, electricity bills etc., 25 per cent on consumer durable goods and 24 per cent is income loss due to requirement of more rest.

Table 9.4: Annualised Averted Expenditures on Consumer Durable Goods Due to Heat Waves

Type of heat wave averted activities at home	Sample proportion who reported to have undertaken these changes due to heat waves	Average cost of the activity or the materials used (in INR) (Alphabets in parenthesis are sources of money)	Possible life span used	Annual average cost of the activity using Eq. 10 (in INR)
Wipe floor repeatedly using cooling ingredients	0.33	320 (S)	One Years	320
Put thick layers of paddy straw on roof to stop transmission of heat	0.54	482 (S+B)	One Years	482
Use thick curtains	0.14	503 (S+B)	5 years	126
Insulate walls with paint, mud	0.017	1020 (S)	10 years	152
Purchase fans	0.92	1295 (S+B)	15 years	151.29
Purchase Air/water cooler	0.05	4233 (S+B)	10 years	631
Purchase fridge	0.04	7162 (S+B)	20 years	729.47
Purchase Air Conditioner	0.003	18000 (B)	20 years	1833.34
Weighted annual average for a sample household				592

Note: S: Saving; B: Borrowing

Conclusion

One form of increased temperature due to climate change is a heat wave, which is a continual spell of hot and humid weather for two or three days, and can be lethal (Basu and Jonathan 2002). High temperature impacts economic output through various channels like effect on health, labour productivity, productivity of crops, labour supply, etc. This paper looked into how heat waves affected time allocation between work and rest by altering the marginal utility from time spent on these activities to workers in occupations that expose them to extreme weather. Low-income category urban workers were surveyed in two different cities of eastern Indian state of Odisha, Bhubaneswar, and Sambalpur, where heat waves have occurred regularly during the summer since 1998. The results show that the working class has been seriously impacted by heat waves; they are not able to work as long as they do on a normal summer day and have to spend more on food, electricity, etc. Other health effects have necessitated hospitalisation in some cases.

The results show that workers work 1.19 hours less and spend 0.46 hours less at home, and they rest 1.65 hours longer on average on a heat wave day than on a normal summer day. Work time loss is more for people doing manual work. Other than work time loss, people are found to be spending extra on routine items like food, electricity bills, etc. and undertaking other adaptive measures to cool floor and roof of the house, using thick curtains made of locally grown grasses, buying fans, water coolers, etc. Annualising this expenditure and adding the extra expenditure on routine goods and income loss due to work time loss, per head private adaptation cost of low-income informal sector urban worker is calculated to be INR195 per heat wave day or INR2340 per annum which constitutes 2.7 per cent of their income. Of this amount, 51 per cent is due to extra expenditure on food, electricity bills etc., 25 per cent due to annualised consumer durable expenditures and 24 per cent due to income loss from more rest. As per Government of India estimates, there were 644000 urban informal sector non-agricultural workers in Odisha in the year 2004-05of which nearly 128000 (20 per cent) work in construction, trade and transport sectors to which the present sample mostly belonged to.[11] Extrapolating the private adaptation cost for this group, it estimates to be INR25.12 million per heat wave day or INR3.01billion per annum at 2012-13 prices for the state of Odisha. Using a simple linear extrapolation, one finds the number of heat wave days to increase 2.5 times by 2020 and 3.5 times by 2030, which will put heavy burden on these people. This also means that, in general, labour markets in low-income countries are likely to be seriously affected in terms of both labour supply and wellbeing of the labour class. As more than 50 per cent of the adaptation cost is on food and other routine expenditure, government intervention – in the form of income supplements, subsidised ration, provision of subsidised electricity or increased wages in summer – can help workers cope better with such climate stress.

cknowledgements

nancial assistance from IDRC Think Tank Initiative (TTI) Seed Grant Scheme
the Institute of Economic Growth (IEG) is sincerely acknowledged. Sincere
anks go to Prof. Mendelsohn, Editor, Climate Change Economics Journal for his
rmission to republish the paper with modification. I also thank A.Mitra, S. Kar,
. Haque and participants of 'Climate change and inequality' conference jointly
ganised by CODESRIA, CLACSO and IDEAS at Dakar, Senegal during June
)14 for useful suggestions. Both Ruchika Khanna and Chandan Jain are sincerely
knowledged and thanked for providing diligent research assistance and S R
Iania and Bijay Behera, for their help in conducting the surveys in Bhubaneswar
d Sambalpur cities respectively.

Jotes

1. JEL classification: J22, J28, Q54, Q58 A more theoretical and extended version
   of the paper has been published in Climate Change Economics Journal (Das, S,
   'Temperature increase, labour supply and cost of adaptation in developing economies:
   Evidence on urban workers in informal sectors' Climate Change Economics, Vol. 6,
   No. 2 (2015) 1550007 (24 pages)).

2. Hyperthermia is caused if the core body temperature attains lethal values like 42–43° C.

3. Certain activities are constrained to work within a certain time frame; for example,
   after the market shuts in the evening, petty traders cannot operate.

4. http://www.osdma.org/ViewDetails.aspx?vchglinkid=GL002&vchplinkid=PL008,
   and http://www.osdma.org/ViewDetails.aspx?vchglinkid=GL003&vchplinkid=PL0
   13&vchslinkid=SL005&vchtlinkid=TL003, both accessed on 21st April 2014.

5. For details, see the IOM website: http://www.iom.int/cms/envmig; ADB paper:
   http://www.adb.org/sites/default/files/pub/2012/addressing-climate-change-
   migration.pdf (accessed on 21st April 2014); UNESCO paper: http://www.unesco.
   org/new/en/social-and-human-sciences/themes/sv/news/migration_and_climate_
   change_a_unesco_publication_on_one_of_the_greatest_challenges_facing_our_
   time/#.UhXh0NKBl6o (accessed on 22nd April 2014)

6. Other comprehensive studies and meta-analyses examine the ergonomics and
   physiology of thermal stress in humans (Pilcher *et al.* 2002; Hancock and Vasmatzidis
   2003; Hancock *et al.* 2007).

7. The exchange rate was US$1 = INR60 during the survey period, i.e. during April-
   May 2013.

8. Nobody reported receiving information from more than three sources.

9. Many reported different causes for having consumer durables like they had money to
   buy, received these as gift, received as dowry from in-laws, etc.

0. http://www.nachi.org/life-expectancy.htm, accessed on 12th January 2015

1. http://dcmsme.gov.in/Report_Statistical_Issues_Informal_Economy.pdf,   accessed
   on 10th February 2014.

References

Becker, G. S., 1965, 'A Theory of the Allocation of Time', *The Economic Journal*, Vol. 7
No. 299, pp. 493–517.

Basu, R. and M. Samet Jonathan, 2002, 'Relation between Elevated Ambient Temperatu
and Mortality: A Review of the Epidemiologic Evidence', *Epidemiologic Reviews*, Vc
24, No. 2, pp. 190–202.

Bynum, G.O., K.B. Pandolf, W. H. Schuette, R. F. Goldman, D. E. Lees, J. Whang-Pen
E. R. Atkinson, and J. M. Bull, 1978, 'Induced Hyperthermia in Sedated Humar
and the Concept of Critical Thermal Maximum.' *American Journal of Physiolo*
Regulatory Integrative Comp Physiology (PLS SUPPLY DETAILS IN FULL) 23
228–36.

Currie, J. and B. C. Madrian. 1999. 'Health, Health Insurance, and the Labor Marke
n *Handbook of Labor Economics,* Volume 3, edited by Orley Ashenfelter and Davi
Card, 3309-416. Amsterdam:Elsevier Science.

Das, S. and S. C. Smith, 2012, 'Awareness as an Adaptation Strategy for Reducir
Mortality from Heat Waves: Evidence from a Disaster Risk Management Progra
in India.' *Climate Change Economics*, Vol. 3, No. 2, pp. 1250010-01 - 1250010-2
(29 pages):

Das, S. 2014. 'Media Use in Public Health Communication on Heat Waves: Examinir
the Effective Medium', *SANDEE Working Paper* (forthcoming).

Dau-Schmidt, K. G.,1984, *The effect of Consumption Commitments on Labour Suppl*
Thesis/dissertation : Microfilm Archival Material .

Epstein Y., G. Keren, J. Moisseiev, O. Gasko and S. Yachin, 1980, 'Psychomotc
Deterioration during Exposure to Heat.' *Aviat Space Environ Med.*, Vol. 51, No. (
pp. 607–10.

Galloway, S.D. and R.J. Maughan, 1997, 'Effects of Ambient Temperature on th
Capacity to Perform Prolonged Cycle Exercise in Man.' *Med Sci Sports Exerc.*, Vc
29, No. 9, pp. 1240–49.

González-Alonso, José, Christina Teller, Signe L. Andersen, Frank B. Jensen, Tino Hyldi
and Bodil Nielsen, 1999, 'Influence of Body Temperature on the Development o
Fatigue during Prolonged Exercise in the Heat.', *Journal of Applied Physiology*, Vo
86, pp. 1032–39.

Gronau, R., 1987, 'Home Production – A Survey', in O. Ashenfelter & R. Layard, ed
Handbook of Labor Economics, edition 1, volume 1, chapter 4, pages 273-304 Elsevie

Hancock, P. A. and I. Vasmatzidis, 2003, 'Effects of Heat Stress on Cognitive Performanc
The Current State of Knowledge.' *International Journal of Hyperthermia*, Vol. 19, 3
pp. 355-372.

Hancock, P.A., J.M. Ross and J.L. Szalma, 2007, 'A Meta-Analysis of Performanc
Response under Thermal Stressors', *Hum Factors*, Vol. 49, No. 5, pp. 851–77.

Heal, Geoffrey M. and Jisung Park, 2014, 'Feeling the Heat: Temperature, Physiolog
& the Wealth of Nations.' *Discussion Paper* 2014-51. Cambridge, Mass.: Harvar
Environmental Economics Program.

IPCC, 2014, 'Summary for policymakers', in: *Climate Change 2014: Impacts, Adaptatior
and Vulnerability*. Part A: Global and Sectoral Aspects. Contribution of Workin

Group II to the Fifth Assessment Report of the Intergovernmental Panel on Climate Change [Field, C.B., V.R. Barros, D.J. Dokken, K.J. Mach, M.D. Mastrandrea, T.E. Bilir, M. Chatterjee, K.L. Ebi, Y.O. Estrada, R.C. Genova, B. Girma, E.S. Kissel, A.N. Levy, S. MacCracken, P.R. Mastrandrea, and L.L.White (eds.)]. Cambridge University Press, Cambridge, United Kingdom and New York, NY, USA, pp. 1-32.

IPCC, 2007a, 'Summary for Policy Makers', in *Climate Change 2007: The Physical Science Basis*. Contribution of Working Group 1 to the Forth Assessment Report of the Intergovernmental Panel on Climate Change, S. Solomon, D. Qin, M. Manning, Z. Chen, M. Marquis, K.B. Averyt, M. Tignor and H.L. Miller, eds, Cambridge: Cambridge University Press.

IPCC, 2007b, *Climate Change 2007: Synthesis Report*. Contribution of Working Group I, II and III to the Forth Assessment Report of the Intergovernmental Panel on Climate Change, edited by R.K. Pachauri and A. Reisinger, 104 pp. Geneva: IPCC.

Kovats R. S. and S. Hajat, 2008, 'Heat Stress and Public Health: A Critical Review.' *Annual Review of Public Health*. Vol. 29, pp. 41–55. http://www.annualreviews.org/doi/pdf/10.1146/annurev.publhealth.29.020907.090843. Accessed on 22 August 2013.

Lan, Li, Pawel Wargocki and Zhiwei Lian, 2010, 'Optimal Thermal Environment Improves Performance of Office Work'. Available at http://www.rehva.eu/index.php?id=151. Accessed on 14 February 2014.

Ma, Y., B.C. Olendzki, W. Li, A.R. Hafner, D. Chiriboga, J.R. Hebert, M. Campbell, M. Sarnie and I.S. Ockene, 2006, 'Seasonal Variation in Food Intake, Physical Activity, and Body Weight in a Predominantly Overweight Population.' *European Journal of Clinical Nutrition*, Vol. 60, pp. 519–28.

MaCurdy, T.E., 1981, 'An Empirical Model of Labor Supply in a Life-Cycle Setting', *Journal of Political Economy*, Vol. 89, pp. 1059-85

Meehl, G.A. and C. Tebaldi, 2004, 'More Intense, More Frequent, and Longer Lasting Heat Waves in the 21st Century.' Science, *Vol*. 305, No. 5686, pp. 994–97.

Mehnert P., J. Malchaire, B. Kampmann, A. Piette, B. Griefahn and H. Gebhardt, 2000, 'Prediction of the Average Skin Temperature in Warm and Hot Environments.' *European Journal of Applied Physiology*, Vol. 82, pp. 52–60.

Mohamed, S. and K. Srinavin, 2005, 'Forecasting Labor Productivity Changes in Construction using the PMV Index.' *International Journal of Industrial Ergonomics*, Vol. 35, No. 4, pp. 345–51.

Nielsen, B., J.R. Hales, S. Strange, N.J. Christensen, J. Warberg and B. Saltin, 1993, 'Human Circulatory and Thermoregulatory Adaptations with Heat Acclimation and Exercise in a Hot, Dry Environment', *Journal of Physiology*, Vol. 460, pp. 467–85.

Niemela, R., M. Hannula, S. Rautio, K. Reijula and J. Railio, 2002, 'The Effect of Air Temperature on Labour Productivity in Call Centres-A Case Study.' *Energy and Buildings*, Vol. 34, No. 8, pp. 759–64.

Pilcher J.J., E. Nadler and C. Busch, 2002, 'Effects of Hot and Cold Temperature Exposure on Performance: A Meta-analytic Review', *Ergonomics*, Vol. 45, No. 10, pp. 682–98.

Pivarnik, J. M., M. J. Reeves and A.P. Rafferty, 2003, 'Seasonal Variation in Adult Leisure-time Physical Activity', *Medical Science Sports Exercise*, Vol. 35, pp. 1004–08.

Ramsey, J. D., 1995, 'Task Performance in Heat: A Review', *Ergonomics*, Vol. 38, No. 1, pp. 154–65.

Robinson, P. J., 2001, 'On the Definition of a Heat Wave', *Journal of Applied Meteorology* (American Meteorological Society), Vol. 40, No. 4, pp. 762–75.

Sherwood, S.C. and M. Huber, 2010, 'An Adaptability Limit to Climate Change due to Heat Stress', *Proceedings of the National Academy of Sciences,* Vol. 107, No. 21, pp. 9552–555.

U.S. Department of Health and Human Services, 1996, 'Physical Activity and Health: A Report of the Surgeon General', U.S. Department of Health and Human Services, Centers for Disease Control and Prevention. Atlanta: National Center for Chronic Disease Prevention and Health Promotion.

Zivin, J.G. and M.J. Neidell, 2010, 'Temperature and the Allocation of Time: Implications for Climate Change', *NBER Working Paper Series*, 15717. Available at http://www.nber.org/papers/w15717.

Appendix

Dos and Don'ts of Heat Wave Awareness Programme of Odisha Government

Precautions to avoid heat stroke

- Eat enough food and drink enough water before going out.
- Consume different types of liquids like water rice, *belapana, sarbat* (locally available sweet drinks), curd water, ORT solution, watermelon, cucumber etc.
- Carry required amount of water if going out.
- Wear light coloured cotton cloths
- Either avoiding travel during noon or use umbrella, cap, turban, wet towel, shoes and if possible, goggles when walking in the sun.
- If too hot, reschedule your work so as to work more during morning and afternoon and less during noon.
- Remain alert for children, elderly, fat people, pregnant women and persons with high blood pressure, diabetes or epilepsy.
- Do not give water if the person faints due to heat attack.
- Do not take alcoholic drinks.
- Please consult doctor if feeling uncomfortable due to heat.

Symptoms of heat stroke (take the person to a hospital if any of these symptoms are found)

- Feeling of tired.
- Headache, body ache and vomiting.
- Dry throat.
- Blink vision
- Abnormal increase in body temperature.
- Increased palpitation
- Being unconscious

www.ingramcontent.com/pod-product-compliance
Lightning Source LLC
Chambersburg PA
CBHW050652280326
41932CB00015B/2874